Constructing Risk and Safety in Technological Practice

Risks and uncertainties are a pervasive aspect of daily life in modern societies. However, it is also clear that technologies and their "risks" have different meanings for the various groups of actors who develop, use or contest them at specific sites. What is viewed as a risk, how risks should be handled and how safety can be achieved are highly contested issues with strong political implications.

Constructing Risk and Safety in Technological Practice represents a new area of social science inquiry in this field where the focus is on understanding how risks are interpreted, handled and debated by individuals or groups of actors in their everyday practice. This book investigates how risks and uncertainties are constructed and negotiated by engineers, managers, hospital workers, laypersons and others in their day-to-day interactions. It considers how conflicts and power relations are acted out in these processes and how safety is achieved. The essays focus upon detailed case studies at specific sites such as engineering companies, intensive care units, nuclear waste handling facilities and air traffic control centers.

The book's multidisciplinary approach will be of interest to a wide range of readers: social scientists, risk analysts and managers in organizations and workplaces, and will also be valuable reading for graduate courses in sociology, management and organization studies.

Jane Summerton, sociologist of technology, is associate professor at the Department of Technology and Social Change, Linköping University, Sweden. Her previous work has focused on managerial and user practices in sociotechnical networks and systems, particularly in relation to energy and environmental issues. Her current research concerns "non-user" practices in technology and their intersections with socio-economic resources, ethnicity and gender.

Boel Berner, sociologist and historian, is professor at the Department of Technology and Social Change, Linköping University, Sweden. Her research has focused on the identities and practices of technical experts, in work, education and in relation to risk, and on questions of gender and technology. Her most recent books in English are *Gendered Practices* (ed. 1996) and *Manoeuvring in an Environment of Uncertainty* (ed. with Per Trulsson 2000).

Routledge advances in sociology

This series aims to present cutting-edge developments and debates within the field of sociology. It will provide a broad range of case studies and the latest theoretical perspectives, while covering a variety of topics, theories and issues from around the world. It is not confined to any particular school of thought.

Constructing Risk and Safety in Technological Practice

Edited by Jane Summerton and
Boel Berner

Routledge
Taylor & Francis Group

LONDON AND NEW YORK

First published 2003 by Routledge

2 Park Square, Milton Park, Abingdon, Oxfordshire OX14 4RN
52 Vanderbilt Avenue, New York, NY 10017

Routledge is an imprint of the Taylor & Francis Group, an informa business

First issued in paperback 2020

Typeset in Garamond by
Florence Production Ltd, Stoodleigh, Devon

British Library Cataloguing in Publication Data
A catalogue record for this book is available
from the British Library

Library of Congress Cataloging in Publication Data
Constructing risk and safety in technological practice/edited by
 Jane Summerton and Boel Berner.
  p. cm.
 Includes bibliographical references and index.
 1. Industrial safety. 2. Risk management. I. Summerton, Jane.
 II. Berner, Boel.
 T55 .C663 2002
 302′.12–dc21   2002068173

ISBN 978-0-415-28571-1 (hbk)
ISBN 978-0-367-60468-4 (pbk)

Contents

Contributors

Tomas Backström is an engineer and sociologist and senior researcher at the National Institute for Working Life in Stockholm. He is also affiliated with the Department of Industrial Economics and Management at the Royal Institute of Technology, Stockholm. His research deals with complex organizations and organizational development, and his most recent projects concern sustainable work systems. He has also numerous publications within the field of accident and safety research.

Boel Berner is a sociologist and professor at the Department of Technology and Social Change, Linköping University. Her main research areas are: the role of technical experts in social change, gender and technology, and risk in blood transfusion systems. Her recent publications include: *Gendered Practices* (ed.), Almqvist & Wiksell International 1997; *Perpetuum Mobile? The Challenge of Technology and the Course of History* (in Swedish), Arkiv 1999; *Suède: l'égalité des sexes en question* (ed. with Elisabeth Elgán), l'Harmattan 2000; *Manoeuvring in an Environment of Uncertainty* (ed. with Per Trulsson), Ashgate 2000.

Sidney W.A. Dekker received his PhD in Industrial and Systems Engineering from Ohio State University and is currently associate professor at the Linköping Institute of Technology in Sweden. Previously with the British Aerospace Dependable Computing Systems Centre in the UK, he has worked in the Netherlands, Australia, New Zealand and Singapore. As a pilot, he has been through first officer training on the DC-9. He is a full member of the Human Factors/Ergonomics Society and the European Association of Cognitive Ergonomics. His latest books are *Coping with Computers in the Cockpit* (Ashgate 1999) and *The Field Guide to Human Error Investigations* (Cranfield University Press 2002).

Marianne Döös is an educational psychologist and senior researcher at the National Institute for Working Life, Sweden. She is also affiliated with the Department of Education at Stockholm University. Her research concerns the processes of experiential learning in contemporary settings. Recent projects concern conditions for learning and organizational development and change. She also has numerous publications within the field of accident and safety research.

Judith Green is a senior lecturer in sociology at the London School of Hygiene and Tropical Medicine, where she teaches methodology and the sociology of health. Her research interests include the sociology of accidents, the organization of health care and public understanding of health. Her recent publications include *Risk and Misfortune: the social construction of accidents* (UCL Press 1997) and *Analysing Health Policy: a sociological approach* (co-authored with Nicki Thorogood, Longman 1998).

Andrew Koehler completed his dissertation entitled *Design for a Hostile Environment: technical policymaking and system creation* at the University of California, Berkeley's Goldman School of Public Policy in 2001. He presently holds a post-doctorial position at Los Alamos National Laboratory, New Mexico, where he is working on technology analysis, with particular focus on the use of agent models to simulate regulatory behavior.

Jörg Potthast is a doctoral candidate at the Wissenschaftszentrum Berlin für Sozialforschung (WZB), where he is currently working on a sociological dissertation on maintenance services at hub airports. He has been a visiting student at the Centre d'Etudes des Techniques, des Connaissances et des Pratiques (CETCOPRA), Université Paris 1, and at the Complex Product Systems Innovation Centre (CoPS), University of Brighton/Sussex. His research interests focus on intersections of science and technology studies and the sociology of work.

Gene I. Rochlin is professor at the Energy and Resources Group, University of California, Berkeley. His research interests include science, technology and society, cultural and cognitive studies of technical operations, the politics and policy of energy and environmental matters, and the broader cultural, organizational and social implications and consequences of technology – including large technical systems. His recent book *Trapped in the Net: the unanticipated consequences of computerization* (Princeton 1997) won the 1999 Don K. Price Award of the Science, Technology and Environmental Politics Section of the American Political Science Association.

Johan M. Sanne was trained in Economics and Human Geography before receiving his PhD in Technology and Social Change. He is currently assistant professor at the Department of Technology and Social Change, Linköping University. His research interests are: the construction of risk, meaning and identity and risk handling and interpretation; the understanding and management of complex sociotechnical systems in different professional communities; and social and cognitive work in sociotechnical systems. He is currently working on a project about learning through the use of simulations in large development projects. Recent publications include *Creating Safety in Air Traffic Control* (Arkiv 1999).

Jane Summerton is a sociologist and associate professor at the Department of Technology and Social Change, Linköping University, as well as Director of Social Science for Sweden's multi-university Energy Systems Graduate Program. In fall 1998 and spring 2001 she also taught (with Gene I. Rochlin) graduate courses in science and technology studies at the University of California, Berkeley. Her research interests are: actor dynamics in techno-scientific networks and systems, especially energy systems; managerial exclusions and user practices in infrastructures; social science theory on technoscience. Her publications include several books as well as a large number of articles in international journals and edited volumes.

Sabrina Thelander completed her dissertation entitled *Bringing Back to Life: creating safety in cardiac intensive care* at the Department of Technology and Social Change, Linköping University, in 2001. She is currently a researcher at that department as well as Director of the Center for Man-Technology-Society. Her research interests include the sociology of work and technology; interactions of medicine, risk and safety.

Acknowledgments

This volume evolved from a workshop entitled "Social Construction of Risk and Safety" that took place in southern Sweden in March 2000. The purpose of the workshop was to bring together social scientists from various disciplines to discuss theoretical and empirical work on how risks are defined and handled at everyday sites of technological practice. The workshop, which was organized by the Department of Technology and Social Change at Linköping University, attracted over thirty invited participants from seven countries. We thank all the participants at this workshop for their lively contributions to our many discussions. In particular we thank our co-organizer of the workshop, Johan M. Sanne, as well as Margareta Norstad and Elin Bommenel for valuable organizing assistance.

We are indebted to a number of colleagues who have read and commented on various versions of our introduction and the project in general. In particular we thank our colleagues within the research group "Technology – Practice – Identity" at the Department of Technology and Social Change, especially Jonas Anshelm, Christer Eldh and Johan M. Sanne, for their numerous insightful suggestions. We also thank Bengt Olle Bengtsson for his constructive comments.

Corinna Kruse of the Department of Technology and Social Change has provided invaluable editorial assistance in preparing the book into publishable form and helping us to maintain a semblance of order among the myriad of texts, authors and lists that went into making this volume.

The workshop and this volume have been made possible through generous financial support from the Swedish National Energy Administration; the Swedish Agency for Innovation Systems, VINNOVA (formerly the Swedish Council for Planning and Coordination of Research, FRN); and the Swedish Council for Working Life and Social Research, FAS (formerly the Swedish Council for Work Life Research, RALF). In addition, the Department of Technology and Social Change at Linköping University has made resources available to support the work.

Jane Summerton and Boel Berner
Linköping, Sweden
April 2002

Constructing risk and safety in technological practice

An introduction

Jane Summerton and Boel Berner

On a stormy night in September 1994, the passenger ferry *Estonia* – a large, seaworthy vessel that was fully equipped with sophisticated technology – capsized and sank in the dark waters of the Baltic Sea on her way from Tallinn, Estonia, to Stockholm. Over 800 men, women and children, many of whom were asleep in their berths in the recesses of the boat and unaware of the catastrophe that was about to befall them, followed her into the sea. Although largely unpublicized outside of Europe, the sinking of the *Estonia* was an unprecedented tragedy in the Nordic and Baltic countries and one of the worst naval disasters in the history of peacetime Europe. It was both a transcultural catastrophe and a social disaster on local levels: many of the passengers who died were parts of groups from the same local community or workplace – they were "lunchroom ladies" taking a break from school cafeterias, office workers on their annual outing and elderly people enjoying the leisure of a post-summer ocean cruise. Almost everyone in the two countries that were worst hit by the accident, Sweden and Estonia, knows someone who perished in the Baltic Sea that night.

Controversy immediately broke out concerning what was the cause of the accident, who was to blame, whether known risks had been adequately dealt with, whether the tragedy could have been prevented through technological, organizational or political means and – not least – whether the rescue operation at sea could have been organized more efficiently and thereby saved more lives. Ultimately, the official report of the investigating committee assigned faulty construction on the part of the German shipbuilding company that had designed the *Estonia*, pointing to a crucial locking mechanism on the ferry's back car-loading panel that had failed to withstand the pressures of a stormy sea. Other contributing factors were a lack of organizational procedures to effectively learn from previous incidents, design problems and failure on the part of the ship's crew to identify and react to the unfolding events in time (The joint accident investigation commission of Estonia, Finland and Sweden 1997).

Meanwhile, investigating journalists pointed to a number of other factors behind the catastrophe, including metal fatigue, deficient design, bad maintenance, bad seamanship, greed and an insufficient rescue organization

(Jörle and Hellberg 1996). The ensuing debates about what actually happened, how the aftermath should have been handled and how the sequences of events should be interpreted have been numerous and bitter. The multiple scenarios that have been presented are highly contested; powerful actors and organizations are alternately blamed or exonerated (Forstorp 1998). Few of the "facts" of the case are unambiguous, stable or accepted by all.

The *Estonia* disaster and the controversies that emerged in its aftermath exemplify the themes of this book in several ways. The catastrophe provides a vivid illustration of the hazards involved in the use of modern technologies and of the difficulties of identifying and handling technological risks. It is yet another tragic example of the problems that are involved when distributing responsibility in accident analysis – that is, after-the-fact analysis that is separated in time and circumstance from the uncertainties, chaos and confusion in which the accident was embedded. Finally, the *Estonia* disaster shows the conflicts that typically unfold when individuals, groups and organizations with varying perspectives and interests in relation to the technologies at hand construct the risks of these technologies in highly different ways.

Other examples of technological hazards and uncertainties abound in modern, Western societies. At the time of writing (winter 2002), three Swedish local communities are in the midst of debating the possible risks and advantages of being "host communities" for storing extensive deposits of radioactive nuclear waste that must be sealed off from intrusion and contamination for the next several thousand years. If these communities accept the roles of being candidates for the selection of one final site, the nuclear industry will initiate a series of complex technical procedures to evaluate the respective sites. These procedures include boring test holes far into the ground, checking for fissures in subsurface rock, measuring ground water flows and recording seismic activity – all of which are prerequisites for safe transport and storage of spent nuclear fuel in a high-security disposal site. In 2006, the final decision will be made as to which community among the three candidate sites will be given responsibility for final storage of the highly radioactive waste. Issues of risk and the handling of risk are centrally important not only for those designing and testing the various technical procedures to be used, but also for the regulatory authorities, local elected officials and – perhaps most of all – local residents. What constitutes a "safe" deposit and why? Who can be trusted – today, and in a future that extends 1,000 years in time? Finally, from a social science research perspective, what interpretations of risk are constructed – and contested – within the various communities of practice involved?

As the examples of *Estonia* and long-term storage of nuclear waste indicate, modern technological systems entail diverse and often frightening uncertainties and risks. In some cases, multiple errors and faulty systems may lead to large-scale disasters, as exemplified by the Chernobyl and Bhopal catastrophes in the 1980s. Other technologies involve risks of a less tangible nature, for example concerning the effects of toxic emissions from community dumps, factories and power plants. In recent years, highly publicized controversies have emerged in relation to such disparate topics as genetically modified foods, BSE, global warming and nuclear power. Even "minor" risks and incidents that stem from technology seem to be pervasive aspects of everyday life. Despite this societal embeddedness, the responsibility for assessing and handling these risks is often highly individualized: for example, everyone now seems to be individually accountable for taking his/her own precautions to avoid eating the wrong foods, making sure to wear a helmet while bicycling or securing one's seat belt. As a result, each of us is more or less continuously faced with the task of interpreting and dealing with numerous risks of varying dimensions.

In light of these dynamics, it is not surprising that risks and their handling have come to be placed high on the agendas of concern of lay people, groups of professionals and policy makers. What constitutes a "risk," how should accidents be interpreted and how can safety be achieved? These questions are often highly contested issues, as illustrated in the extensive debates and controversies that often erupt over the failure of specific technological systems or artifacts (for example, *Estonia*'s faulty locking mechanism). It is also clear that the processes of defining risks and the outcomes of these processes – such as delegating responsibility for doing something (or not having done something) about these risks – have strong political implications. Such issues will be explored in this volume.

More specifically, this book focuses on the following questions:

- How are technological risks and uncertainties constructed, negotiated and handled by different actors in their everyday interactions and practices?
- How and why do actors' interpretations of accidents and uncertainties become stabilized as accepted "facts," and how can we understand these processes and outcomes?
- By what means is safety achieved by operators, workers and laypersons in various types of organizations and complex technological systems?

Our perspective on risk will be outlined in this introductory chapter. Succinctly expressed, this book focuses on ongoing practices, situated experiences and socially embedded interactions among actors in processes of constructing risk. We discuss how risks and uncertainties are defined, accidents interpreted and safety achieved by individuals, groups and organizations in everyday, concrete practices of daily life. And we focus on interactions at specific sites – workplaces such as engineering companies, intensive care units, air traffic control centers and meeting rooms.

In this Introduction we will relate our perspective to selected well-known approaches to risk within the social sciences. We will discuss relevant work within several fields, mostly notably cultural and sociological studies of risk

– even though we are aware of the fluidity of boundaries between some work in these areas. We will emphasize the insights from the various studies that we find particularly useful, but we will also critically discuss their limitations from the perspective of this book. We will then focus on two areas of research within sociological studies of risk which seem particularly useful in understanding how risk and safety are constructed in interactive processes and practices, namely (1) studies of how risks, accidents and safety are constructed in organizations and complex sociotechnical systems and (2) studies of how risks are constructed in conflicting interactions among social groups.

Our ambition in this Introduction is thus not only to present our perspective but also to provide a selective view of contemporary social science work on risk issues. We will focus primarily on work that in various ways explores the ways in which actors and groups construct risks and safety at everyday sites of sociotechnical practice. In this discussion, we will continually strive to relate the various perspectives to each other. By identifying both common themes and important differences, we also hope to demonstrate the interweavings between some of the approaches that we have chosen to refer to as "cultural" or "sociological." We hope that the text can serve as a useful introduction for students and researchers interested in exploring this multifaceted work.

Theoretical approaches to processes of constructing risk, accidents and safety

As Freudenburg and Pastor (1992: 389) remind us, risk "is a concept coming of age." Hardly existing in the 1970s, the analysis of risk has become a pervasive, ongoing and institutionalized part of modern, Western societies. It is clear that a multitude of actors and organizations – commissions on national and international levels, safety experts, public authorities, local action groups, insurance companies and researchers in a multitude of fields – are engaged in efforts to interpret and assess the risks and uncertainties of modern technological life (Green 1997; Green this volume).

Theoretical approaches to studying risk and uncertainty can be identified and distinguished in a number of useful ways (see Bradbury 1989; Haukelid 1999; Lupton 1999; Nelkin 1989; Renn 1992; Rosa 1998). Jasanoff, for example, discusses the "two cultures" of risk analysis, namely quantitative, model- and measurement-oriented analyses on the one hand, and qualitative studies on the ethnical, legal, political and cultural aspects of risk, on the other hand (Jasanoff 1993). Jasanoff notes that substantial progress has been made in bridging these two cultures, which now share numerous assumptions with regard to such aspects as the intertwining of facts and values in issues of high uncertainty, and the role of cultural factors in shaping the way people assess risk. Jasanoff's analysis points to a number of potentially fruitful ways in which qualitative researchers can continue to contribute to the fields of qualitative risk analysis.

In the following sections, we will critically explore examples of work that share certain features with our orientation. We will first briefly discuss relevant insights from certain psychological approaches to risk before focusing on work from cultural and sociological perspectives.

Psychological studies of risk

Among psychological approaches to risk, many experimental analyses and surveys have studied how people estimate risks and make choices among various problematic alternatives. Questions asked in this kind of research are, for example: How do people weigh the social benefits as opposed to the social costs of technological change? How do they perceive risk and respond to information about possible dangers, and how are responses to a "risk event" amplified or attenuated? What factors lie behind different kinds of responses to technological risks, hazards and accidents?

An important focus in these studies is to understand why people consider certain risks more important than others or why they trust certain technologies more than others. One finding of this work is that when assessing the costs and benefits of particular risks, individuals generally do not use the type of technical probability analyses favored by economic and technical professionals. In the face of uncertain information and ambiguous messages, they employ rule-of-thumb heuristics. These interpretations may, for example, lead them to overestimate the frequency of rare but memorable and highly threatening events (such as homicides) and to underestimate the frequency of less publicized but more common ones (such as suicides). These and other results provide useful corrections to an overly rationalistic approach to risk (see, for example, the articles collected in Löfstedt and Frewer 1998). They should in principle also be useful to sociologically oriented approaches to risk, since the sociological literature on risk necessarily includes assumptions about how individuals perceive, react to, and choose among risks. As Heimer argues:

> We cannot have an adequate sociology of risk that discusses how decisions of individuals, organizations and governments fit together if we are sloppy about such central matters as individual perception of risk, risk aversion and risk seeking, and estimation of risk.
>
> (Heimer 1988: 492)

In Heimer's view, therefore, the major results of psychometric research should also be taken seriously by those who study risk from broader sociological perspectives.

There are, however, a number of problems with psychological approaches (including psychometric research) in addressing the issues that are the focus of this book (Freudenburg and Pastor 1992; Heimer 1988; Lupton 2000; Wilkinson 2001). One is the obvious difficulty of transferring experimental results to real-life settings: psychological experiments are artificial situations in which the role of information and context is highly problematic, the tasks are often irrelevant, and the subjects are unfamiliar with the tasks. Also, as Caplan notes, people are seldom isolated in making their decisions around risk but are instead embedded in social environments; they "discuss with relatives, neighbours, friends, colleagues, religious advisers, drawing upon their advice and personal experience" (Caplan 2000: 23). The focus on atomized individuals in experiments and surveys thus limits the usefulness of much psychological research.

A second, related type of critique has pointed out that people normally evaluate risk and make decisions in relation to their whole life situations, an aspect that is not taken into account in most psychological research. Life choices also reflect individuals' positions within organizations and institutions such as the family, workplaces, and religious and social groups, and the social location modifies how the heuristics indicated in experimental work are used – if at all. In addition, it is well known that individual perceptions of risk are also influenced by the arguments and debates concerning hazards that are rhetorically prevalent in a particular society at a certain time. The political dynamics of how risks are communicated in social processes influence what phenomena various groups view as risky, how these risks are formulated, whose voices are heard and whose are silenced.

Finally, many sociologists and others have pointed to fundamental problems with the underlying realist approach to risk that often informs psychometric studies. Risk is largely seen as a taken-for-granted objective phenomenon that can be accurately assessed by experts with the help of scientific methods and calculations. The "risk object," that is the technologies or activities to which social agents ascribe the ability to cause harm (Hilgartner 1992), is accepted as more or less given. The phenomenon to be explained is primarily the malleability of *risk perceptions*, that is: how lay people will perceive, understand and react to these "objective" risks. There is often a distrust of the public's ability to judge risks – or at least a concern that the public will not react "properly" to the "real" risks involved and therefore will not take what some experts consider to be appropriate personal or political action.

Cultural studies of risk

In contrast to psychological perspectives, the point of departure for sociological and cultural studies of risk is the malleability of the *risk objects themselves*. As Clarke and Short remark, such "objects might be dangerous in some objective sense ... but the basic sociological task is to explain how social agents create and use boundaries to demarcate that which is dangerous" (Clarke and Short 1993: 379). What is identified as a risk is not based on an a priori "fact" as identified by technical or medical experts; neither are risks merely rule-of-thumb assessments made by isolated individuals. The

ways in which individuals – including experts – interpret risks can instead
be seen as an expression of socially located beliefs and world views that to a
large extent stem from the individual's situated position and experiences
within social hierarchies, institutions and groups.

This is the position taken by the most well-known cultural approach
to risk as formulated by Douglas, Wildavsky and others (Douglas 1992;
Douglas and Wildavsky 1982; Rayner 1986, 1992; Schwarz and Thompson
1990). This work suggests that social location affects both how fearful people
will be of known risks and what particular phenomena people will find
dangerous. In specific situations and contexts, different people may focus on
different issues in relation to the hazards at hand. To borrow an example from
this volume, what many environmentalists worry about in relation to diesel
fueled cars is the growth of automobilism *per se*, while representatives of the
car industry argue in defense of specific problems such as CO_2 emissions
(Summerton, this volume). Because it is not self-evident that the various risks
can be translated into a common denominator (as is often assumed in both
technical and psychometric risk analysis), conflicts are likely to emerge over
what constitutes a risk and how this risk should be handled.

Douglas and Wildavsky (1982) situate the discussion of risk within the
broader perspective of a cultural relativity as to what is perceived as
dangerous, sinful or hazardous. Social agents continually create boundaries
between themselves and others in order to forge group cohesion and attribute
blame for danger and things gone wrong. While the exact boundaries vary
among groups, there is a social rationale behind what individuals and groups
will view as threats: social positions within organizations and institutions
provide people with classification systems and modes of perception that make
certain dangers appear more relevant than others. In other words, risk is a
collective construct – albeit with all the asymmetries that stem from actors'
differential access to power and authority.

An important theoretical tool for culturalist analysis is the "grid/group
theory" of society as outlined by Douglas and Wildavsky (1982; see also
Douglas 1992, Rayner 1992 and Schwarz and Thompson 1990). According
to grid/group theory, what is conceived of as hazardous, how serious that
risk is thought to be, and how it should be handled will be perceived differently depending upon the group to which a person belongs and identifies
with. On the basis of people's experiences of hierarchical stratification and
constraint (the grid dimension) and experiences of team membership (the
group dimension), four distinct cultural types can be identified, namely hier-
archists, individualists, egalitarians and fatalists. Each group has its specific
cosmology and characteristic attitude to risk.

Rayner (1986) elaborates this point in a study about how radiological
hazards were viewed by medical professionals and support staff in American
hospitals. The way in which various individuals reacted to risk was influenced
in particular by their experiences of hierarchical stratification and their experiences of team membership. Specifically, people who were least constrained

by hierarchy or team membership tended to feel secure; they were the "risk takers" – for example, surgeons, specialists in nuclear medicine and radiotherapists. Others (such as medical administrators and radiotechnicians) were "risk deniers," who tended to downplay the risks of radiation, relying on the bureaucratic measures of safety and radiation control. On the other hand, powerless people with little feeling of team membership (such as maintenance workers and junior nurses) expressed low confidence in the safety measures that were adapted, as well as a high concern for radiation risks.

This kind of cultural approach makes significant contributions to understanding the socially and culturally constituted nature of risk perceptions and, more specifically, how the concrete interpretations of hazards vary among different social groups. It provides important insights into why people will not only react differently to the same hazards but also define hazards differently depending on their social identification and power. This approach has stimulated considerable research both in support of and in opposition to its specific claims.

While we appreciate the emphasis on the socially grounded nature of views and beliefs about risks, we also agree with some of the criticism that has been voiced about some of the studies in this field. The perspective has a tendency to be rigid in view of the multiplicity of groups and perspectives in concrete, everyday situations. The grid/group typology specifies four general outlooks – but why not five or eighteen? In many social contexts, individuals have less clear-cut points of view and may combine outlooks or change perspectives in the course of their lives (or even in different situations of daily life). It is therefore perhaps not surprising that the grid/group typology has failed to predict opinions of risk in several surveys (Sjöberg 1997; see also Bellaby 1990; Boholm 1996).

In our view, this type of approach also has problems capturing the *dynamic, interactive practices* through which actors acquire, construct, negotiate and contest interpretations of their situations and the risks they face in everyday life. (It should be noted, however, that there is a growing body of cultural analysis that focuses upon constructions of risk in dialogues and conversations. This work includes social interactionist approaches to interpreting discourses in the Goffman tradition, see Adelswärd and Sachs 1998; Linell and Sarangi 1998; as well as critical discourse analysis of risks as inspired by Fairclough, see Forstorp 1998.) Also, as Heimer has argued in relation to Rayner's hospital study, this work does not sufficiently emphasize that the various perceptions are also *competing* perceptions:

> When research physicians want to ignore precautions that they believe are unnecessary, they must contend with the health officers responsible for enforcing the regulations; when janitors and maintenance workers want a way to check for themselves whether radioactive materials have been poured down a sink, they have to argue with health officers who

may believe either that they should be the ones to do the checking or that the workers are merely paranoid.

(Heimer 1988: 500)

From such a perspective, interpreting and defining risks involves processes of negotiating among competing views, also handling the conflicts and power relations that are often entailed in such negotiations. These dynamics shape what risks are constructed and by whom, how these risks are framed and whether or not they are acted upon – aspects that a sociocultural theory of risk also must take into account.

Sociological studies of "risk society"

As we noted earlier, many cultural approaches tend to have a transhistorical scope. Douglas does not recognize clear distinctions between traditional and modern societies, instead arguing for continuities between our present culture and that of earlier periods of human history. Accordingly, there is no particular emphasis on the role of modern science or technology in constructing new kinds of hazards and uncertainties.

In contrast, the macro-sociological approach to risk – as presented most notably by Beck and Giddens – situates the perception and handling of risk within analyses of large-scale structural transformations of modern societies. Beck focuses on the relations between modernization, ecological crisis and political processes in Western "risk societies" (1992, 1995; see also Adam *et al.* (eds) 2000) while Giddens analyzes the institutional dimensions of modernity, the concepts of chance and risk, and their significance for individual self-identity (1990, 1991). In these approaches, societal processes in a broad sense provide points of departure for discussing issues of, for example, the spatial extension of global threats, the increasing dependence in modern societies upon experts and expert systems and the challenges of creating trust in post-traditional societies.

Beck in particular emphasizes the role of *modern technologies* in contributing to the risks that threaten us, thus sharing many of the concerns that are the point of departure for the current volume. For Beck, it is precisely the Western technological rationality and scientific institutions that once inspired faith in modernization and progress that have now brought us to the brink of environmental catastrophe. We can no longer insure ourselves against the risks brought about by nuclear, chemical and genetic technologies, and these types of "modernization risks" are historically unique through their global nature and modern causes (Beck 1992). Salvaging the modern project entails making a transition from an industrial modernity with its logic of rationality to a new "reflexive modernization" that is characterized by a critical, reflexive stance to technological risks and the logic that undergirds these risks. Beck notes:

The concept of risk is directly bound to the concept of reflexive modernization. Risks may be defined as a systematic way of dealing with hazards
and insecurities induced and introduced by modernization itself. Risks,
as opposed to older dangers, are consequences which relate to the threatening force of modernization and its globalization of doubt. They are
politically reflexive.

(Beck 1992: 21)

Beck argues that the potential for developing a new form of reflexive
modernity lies in the subpolitics of modern societies, that is, political movements that are outside of the formal, parliamentary-based political system
(such as environmental and peace organizations, social protest movements
and the like). Beck's analysis is thus a powerful critique of modern civilization with its global risks, while also underscoring the urgency of reflexively
dealing with these risks. It has also stimulated a large body of discourse
analytical approaches to understanding risks of modernity, particularly
ecological or environmental risks (see Cohen 2000; Hajer 1995; Lash *et al.*
1996; Mol and Spaargaren 1993).

Beck's critique stands in contrast to Douglas and Wildavsky's position in
important ways. For Douglas and Wildavsky, the recent fears of an ecological catastrophe do not necessarily indicate that new forms of grave dangers
have emerged. Douglas and Wildavsky indeed assert there are real hazards
in the world; Douglas is keen to note, for example, that:

the dangers are only too horribly real . . . this argument is not about the
reality of the dangers, but about how they are politicized . . . Starvation,
blight and famine are perennial threats. It is a bad joke to take this
analysis as hinting that the dangers are imaginary.

(Douglas 1992: 29)

However, Douglas and Wildavsky view today's ecological movement as
expressing a socially located point of view: it reflects a culturally conditioned
way for sectarian groups to express group solidarity and morally legitimate
their opposition towards established power structures. The anti-nuclear, anti-
pollution scenario is particularly suited, they argue, to an egalitarian way of
life and can thus be seen as a device for casting blame upon powerful "others"
who are identified as a threat to the anti-authoritarian lifestyles preferred by
the egalitarians (Douglas and Wildavsky 1982).

Vera-Sanso has recently polemized against Beck in a similar vein, arguing
that rather than any new environmental threats, it is instead uncertainties
about social and economic conditions in the West that have brought about
a kind of millennial *angst*. Vera-Sanso notes that technological risk

becomes the stand-in for distant, unaccountable, often unidentifiable,
shifting "centres" of power. In this context, Dolly the cloned sheep,

Chernobyl, Creutzfeld Jakob's Disease and Thalidomide are metaphors for society's nightmare . . . Having failed to realise the most recent incarnation of the "white man's burden" – global development – and unsure of their own social and economic future, certain sections of the West's population now aspire to save the world from man-made risks.

(Vera-Sanso 2000: 112)

Today's "risk-talk," she argues, must be contextualized as political discourse that reflects the fears and ethos of a particular society, as well as the distribution of power within that society. In Western societies, *social* risks (that is, risk to one's status) have been considerably reduced; therefore, Vera-Sanso argues, risk-talk and attributions of blame have shifted to environmental hazards. In contrast, the most highly charged risk-talk in India attributes risk and blame to socially disruptive individual behavior that is viewed as undermining the social order. More specifically, the everyday risk on which much of contemporary Indian society is predicated is the threat posed to the social order by *women* – the emancipation of whom may lead to broken families, illegitimacy, divorce, abandonment of the elderly and the like (Vera-Sanso 2000: 113).

Beck clearly does not view today's ecological or anti-nuclear concern as just another "special group response" to perennial threats (as in Douglas and Wildavsky's interpretation), or as the expression of a particular society's ethos (as in Vero-Sanso's view). In opposition to the perspective of writers such as Douglas, Beck argues:

whilst correctly appraising the cultural relativity of hazard perception, there is smuggled in an assumption of equivalence between incomparable terms. Suddenly everything is dangerous and thus equally safe, judgeable only within the horizon of its estimation. The central weakness of cultural relativism lies in its failure to apprehend the special socio-historical features of large-scale hazards in developed technological civilization.

(Beck 1995: 96, quoted in Wilkinson 2001: 6)

Earlier we criticized the cultural approach to risk for not sufficiently recognizing the diversity and changing dynamics of risk constructions as they are expressed in the multiple situations and practices of everyday life. A similar critique might also be directed against Beck's approach. More accurately, however, Beck's work reflects an approach to risk that is fundamentally different from the perspective of this book. While Beck insightfully analyzes the risks of modern technological societies within a generalized critique of Western civilization and the rationality that sustains this civilization, we focus on understanding how specific technological risks are interpreted and handled among various groups. Expressed another way, while Beck challenges and rejects the overall rationale of modern, industrial societies, we choose to understand the relevant actors' own, situated rationalities. (As we shall see later in this Introduction, this does not mean that we belittle the challenges that employing modern sociotechnical systems entail, rather that we focus on the interactions, power relations and practices that frame the hazards and uncertainties involved.)

Relating our approach

The perspective of this volume can be summarized in three points. First, while some approaches to risk (such as Beck's work) focus on broader sociocultural patterns, the contributors to this volume focus on *micro- and mesolevel processes* of interpreting and handling risk in organizations and sociotechnical systems. The authors share a concern for understanding the interactive processes by which phenomena such as hazards, risks, uncertainties and safety are constructed in everyday practices.

Related to this point, the second feature that characterizes the contributions in this book is that they are grounded in *extensive empirical work*. As Lupton (2000: 6) has noted, many social science approaches to risk "tend to operate on the level of grand theory with little use of empirical work into the ways that people conceptualize and experience risk as part of their everyday lives." In contrast, the various chapters in this volume are case studies that examine interactions at specific contexts and sites – such as workplaces, air traffic control centers, intensive care units, nuclear waste facilities and other high technological milieus.

Third, the contributors to this book assert and explicate the *constructed nature of risks and hazards*. While (as formulated by Fox 1999) traditional engineering or psychometric analyses held the view that "a risk maps directly on to an underlying hazard," and authors such as Douglas and Beck seem to argue that "hazards are natural, risks are cultural," our interest lies in also exploring how risk preceptions constitute the hazardous – that is, how risks, accidents and safety are constructed as meaningful entities in the interactions and negotiations of everyday life.

Constructing risks in organizations and sociotechnical systems: ongoing practices at everyday sites

Two specific areas of sociological and cultural research seem particularly useful for understanding the interactive practices that are the focus of this book. The first area of research focuses on *organizations and sociotechnical systems* as the locus for creating or avoiding disasters related to advanced technology; the second area of research focuses on constructions of risk in conflictual and dynamic practices within and among different *social groups*.

Accidents and safety in complex organizations and sociotechnical systems

The 1980s witnessed a number of technological disasters in many parts of the world, as well as many analyses of the processes that led to these events.

Examples of well-researched disasters include the sinking of the ferry *Herald of Free Enterprise* outside Zeebrugge, the underground fire at King's Cross Station, the oil rig accidents in the North Sea, the *Challenger* accident and the Bhopal, India, disaster. In Bhopal, at least 4,000 people died, more than 200,000 were injured and hundreds of thousands had to flee their homes when poisonous methal isocyanate (MIC) gas from a Union Carbide plant escaped and contaminated the city (Shrivastava 1987). Shrivastava analyzes the complex, interwoven factors that contributed to the disaster. While management blames one employee for having been careless, Shrivastava points to many organizational, technical and managerial failures that contributed to the catastrophe. Among these factors were poor worker training, poor motivation among workers and managers, insufficient staffing, economic difficulties, discontinuities in top management, unanticipated interactions among equipment, tight coupling of parts, design flaws and defective and malfunctioning equipment (as summarized by Roberts 1990: 104). Neither the workers nor the many poor people who lived in shacks just outside the factory gates were informed about the toxicity of MIC; no alarms went off, no authorities were notified and the evacuation was done in chaos. A number of circumstances, actions and neglections thus combined to produce the catastrophic outcome.

Shrivastava's analysis is an example of the increasing awareness among both researchers and accident investigation commissions of the complex intertwining of technological, organizational, political and social factors in technological failures (see also Reason 1990; Turner 1978). Starting in the 1980s, the previous tendency among investigative commissions in particular to delegate blame to individuals was to a great extent abandoned for a more nuanced, sociotechnical view (although, as the article by Dekker in this volume indicates, this shift is far from universal). The work of Perrow (Perrow 1984; new edition 1999) was instrumental in stimulating this emerging understanding within social science studies of risk. While Shrivastava's analysis focuses on organizational *failure* as a source of risks, Perrow's study, *Normal Accidents*, was an important point of departure for understanding how accidents may occur as part of the *normal* functioning of complex organizations and sociotechnical systems. As Vaughan argues in a 1999 overview, Perrow's work "charged the intellectual community and changed the research agenda. The result is a body of work showing how institutions, organizations, technologies and cognitive practices contribute to accidents and disasters" (Vaughan 1999: 292; see also the Afterword to Perrow 1999).

Perrow focused attention upon the decision-making dilemmas involved in operating high hazard, sociotechnical systems such as nuclear power systems, giant oil tankers and large chemical plants. These systems are operated by highly complex organizations that are characterized by overlapping interactions and tight coupling between their constituent parts. As a result of these complex interactions, incidents are difficult to understand and diagnose, and emergent problems spread quickly throughout the system. On the one hand, these dynamics indicate that those operators who are close to the actual work-ings of the system are crucially important: it is these actors who must make quick decisions in the event of problems. On the other hand, the character-istic tight coupling among components calls for swift, concerted action from those who have an overview of all system operations, namely top management. System characteristics thus call for a *simultaneous* decentraliza-tion and centralization of decision making – a dilemma that is not easily solved. Perrow argues that accidents in such systems are therefore a "normal" outcome – that is, something that can be expected to occur sooner or later. Also, because these systems often handle highly toxic or explosive material, the potential for disasters is omnipresent. Considering the widespread horror and damages that large-scale disasters incur for the workers and the communities affected, Perrow concludes that complex, tightly coupled, high hazard technological systems should either be abandoned, drastically scaled down or drastically redesigned (Perrow 1999: 363ff.).

Perrow's approach is essentially a structural view: he sees accidents as the result of the structural complexity and coupling within a system, and he situates individual judgments and choices within a number of constraints and forces. In Perrow's view, actors simply have no chance to perceive failures and risks in the face of organizational complexity, tight coupling, unanticipated interactions, rapid escalation of events and the like. Certain systems are what he refers to as "error inducing": they are prone to multiple errors that might bring about unexpected interactions, which in turn can undermine safety measures and actors' abilities to manipulate the various factors involved.

Perrow's analyses have been criticized within other areas of organizational research. Specifically, researchers that focus on High Reliability Organizations (HROs) have contested Perrow's arguments (Halpern 1989; Roberts 1990; Rochlin 1991; Weick 1987; see also Rochlin this volume). These researchers have studied precisely the kind of sociotechnical systems that according to Perrow "should" have accidents or near accidents quite often – yet they have found a remarkably safe performance in these systems. How can such safe performance be explained?

HRO researchers have studied a number of complex systems such as nuclear power systems, air traffic control systems, and systems of marine aircraft carriers. These studies point to numerous factors that contribute to why and how these systems are clearly "working in practice but not in theory," to cite two HRO researchers (La Porte and Consolini 1991). In all of these systems, a crucial aspect is actors' abilities to react to unexpected sequences of events, that is, to overcome the decision-making dilemma outlined by Perrow. In their attempts to manage complexity and achieve safety, these organizations

- engage in *constant training* in hopes of "acting out" all possible scenarios that operators could face;

- build in *redundancy* in their operations, that is, employ more people with overlapping positions and competencies, develop technical back-ups and the like;
- are characterized by *flexible organization*, which allows for decentralization of decision making to those near the problems, thus overriding formal hierarchies in specified and hazardous situations; and
- make extensive use of numerous channels of *direct communication* to supplement the many indirect channels that characterize advanced, automated technological systems.

In other words, organizational flexibility and redundancy compensate for actors' limitations and contribute to supporting decision making in the midst of such contingencies as time constraints, stress and high complexity. HRO researchers were among the first to empirically – and in great detail – study everyday practices for handling risks in high hazard organizations and sociotechnical systems. This work has identified a number of concrete, ongoing practices that groups of actors use to assess hazards and create safety in their everyday operations.

There are, however, some concerns about the specificity of the HRO findings and under what conditions these findings might apply (Clarke and Short 1993: 390ff.; Perrow 1999: 369ff.). The organizations studied have rather specific practices that are radically different from those of most other organizations. For instance, the naval vessels that have been analyzed have invoked essentially unlimited control over their operators, who are drilled incessantly and severely sanctioned when they do not assume responsibility. Operators' cognitive premises are set and enforced with a minimum of outside influence; a set of core values is hammered out and socialized into these operators before they are allowed to act independently and autonomously. While some of these conditions can be found in the other HRO organizations studied as well, one can question their political desirability: how many organizations would want to implement such "totalitarian" practices – and should operators accept the uniformity of action and thought thus implied?

Also, HRO research has shown that the *same* aspects of organizations that contribute to their positive handling of incidents (that is, to safety and reliability) also lead to negative behaviors (that is, to mistakes, misconduct and disaster). As Vaughan (1999: 297) points out,

> Training, often used to prevent errors, can create them; information richness introduces inefficiency, too little produces inaccuracy; teams have multiple points of view that enhance safety, but as they become a cohesive group they share assumptions, so the "requisite variety" important to safety is lost.

It is thus still unclear what conditions, practices and contingencies actually contribute to achieving safety in complex organizations and sociotechnical systems – and what factors might instead contribute to increasing the likelihood of accidents. There are also clear differences in orientation between the two approaches of "normal accident theory" and "high reliability theory." While the former studies failures, emphasizes *structures* and argues that certain complex systems will inevitably fail, the latter focuses on safe systems, emphasizes *processes*, and argues for effective prevention (Vaughan 1999: 296f.). New forms of understanding may emerge, Vaughan argues, "when individual projects strive to merge levels of analysis, examining the link between environmental factors [such as political structure, market competition and conditions of war], organizational characteristics, and ethno-cognition in tasks of people doing risky work" (ibid.: 297).

One interesting attempt to analyze such linkages is Vaughan's comprehensive study of the decision to launch the *Challenger* space shuttle in January 1986 – a decision that resulted in the shuttle explosion and the deaths of all on board (Vaughan 1996). Vaughan presents a complex analysis of how factors such as organizational complexity, power relations, "group think" and other situational characteristics of an organization that at the time was considered to be highly reliable (that is, NASA) could contribute to the disaster. Among others, Vaughan's study suggests the importance of viewing complex organizational processes as *arenas* in which judgments, power and passions are played out in specific situations and over time. Within complex organizations, norms and rules for maintaining safe procedures (such as norms with regard to redundancy, training and communication) often express economic and political priorities. Safety is balanced against profit, design change against playing it safe and so on. Official procedures – including those for interpreting what constitutes a safe technology – are modified to fit with emergent practices or with what seems to be economically or politically expedient in a particular situation. This "normalization of deviance" is amply demonstrated in Vaughan's study. Thus Vaughan's study emphasizes the need to study evolving hazards and definitions of risks *in interactions* and *over time*; it also underlines the role of power, authority and contingency in interpreting what constitutes risks and how such risks should be handled.

Constructing risks in dynamic practices among social groups

Power, authority and contingencies are also core issues in many approaches to understanding the practices by which risks and hazards are defined and negotiated in interactive practices among various types of social groups (see, for example, Clark and Short 1993; Freudenburg and Pastor 1992; Hannigan 1995; Lidskog *et al.* 1997; Rosa 1998; Wynne 1989a). Similarly to the work of Douglas, Vera-Sanso and Vaughan, many of these analyses emphasize that beliefs about risks and hazards emerge within particular social contexts and in relation to actors' situated positions in heterogeneous ecologies of knowledge, power and politics. In contrast to the structural determinacy of some approaches within cultural theory, however, the point of departure is that definitions of risk are the *outcomes* of negotiations, controversies and struggles among actors and groups who ascribe diverse meanings to the risks at hand. In these struggles, concepts of risk are often used to legitimize or challenge hierarchies of power.

This perspective is a basic premise of many studies on processes of risk construction within the heterogeneous cluster of fields known as science and technology studies, S&TS (see, for example, Jasanoff *et al.* 1995). Studies within this broad area have focused on political controversies related to risk (Nelkin 1984), the role of science and scientific expertise in these controversies (Wynne 1989a), as well as relationships between constructions of risk, science and policy in regulatory decision making (Jasanoff 1991). Numerous studies have analyzed controversies over siting issues, particularly local struggles among groups over the siting of radioactive nuclear wastes (Lidskog and Litmanen 1997; Sundqvist 2002). There is also a growing body of recent work on *discursive practices*, that is, the rhetorics of risk or "risk-talk" by which groups of actors attempt to sustain, strengthen or legitimize their own knowledge claims while contesting the claims of other groups. In such studies, the empirical focus is typically on the texts, written and verbal exchanges, and public debates that constitute the risk discourse among, for example, groups of scientific experts, activists, policy makers and members of the public(s). Examples of questions that are explored are: how are concepts of risk and uncertainty used as resources to frame and control the discourse surrounding such technologies as nuclear power, genetic testing procedures and food safety? How can the stabilizing of certain meanings of risk be explained? By what means are risks, politics and culture mobilized by specific groups of actors – and what are the consequences for public understanding of these risks?

In contrast to the rather extensive literature on risk discourses, however, relatively little work has been done on the types of interactive practices among groups at specific sites that are an important focus of this volume. In the following we will give two examples of such studies.

In a well-known case study of sheepfarming in northern Cumbria, England, in the aftermath of the Chernobyl nuclear accident, Wynne examines the conflicting ways in which two groups of involved actors interpreted the contaminating impact of radioactive fallout on local soils (Wynne 1989b). Scientists and government officials, on the one hand, and groups of sheepfarmers in northern Cumbria, on the other hand, had radically different views of the extent to which the soil was contaminated and how long this contamination would last. Based on their local practices, extensive knowledge of local soil conditions and certain other factors, the sheepfarmers repeatedly contested the scientists' and officials' assertions about the rate at which the radioactive contamination in the soils actually decayed, claiming that this decay (or detoxification) was not as rapid as the scientists predicted. The scientists, lacking knowledge of the sheepfarmers' local practices and judgments, assumed that their own expert knowledge could be applied without being situated in (or adjusted to) local circumstances such as – notably – actual, local soil conditions. The disparities in the two groups' knowledge claims reflected "a cultural dislocation between two different styles of knowledge; each reflected different kinds of social relations within which knowledge existed" (Wynne 1992: 289). Besides differences in the social relations of knowledge, the hazards that were defined by the two groups also illustrated the many differences in scientists' and sheepfarmers' everyday practices. While the scientific experts' were concerned with standard rules, control and formal knowledge, the sheepfarmers had extensive local knowledge of sheep grazing habits, local soil conditions and actual farming practices.

Wynne argues that conflicts over risks should be understood within "the broader issue of how authority is generated and maintained" (Wynne 1992: 276). Expert knowledge becomes a crucial resource in asserting and maintaining authority, achieving legitimacy and building trust. Far from being "objective" or devoid of social and political considerations, however, expert knowledge and practices are deeply embedded in assumptions, values and social–institutional commitments. These values and commitments frame the way in which scientists and technical experts view risks, their causes and consequences. Wynne shows that how judgments are made – as well as which interpretations become accepted, stabilized and acted upon – is culturally contingent and highly political. This finding is in turn consistent with extensive work within S&TS that focuses on the processes by which "technoscientific" authority is constructed and maintained in specific sites of technological practice (see, for example, Bowker and Star 1999; Knorr-Cetina 1999; Latour 1987, 1999).

A second example of empirical work on interactive practices among groups in constructing risks focuses on processes of interpreting a technological catastrophe. Clarke analyzes an accident investigation concerning the explosion and chemical contamination of an eighteen-story office building in Binghamton, New York, in which 500 people were exposed to toxic chemicals (Clarke 1989). Contrary to a common assumption that this kind of accident is handled efficiently and appropriately, the muddled and chaotic response by the involved organizations and state agencies instead accentuated the crisis. In an ambiguous situation in which neither the technologies nor their hazards were clearly understood, the actors had difficulties utilizing organizational resources and making political decisions.

Clarke outlines the struggle between various groups and organizations to "capture" the right to define the accident. Ultimately, state organizations – rather than other possible actors such as local health departments or environmental organizations – managed to define themselves as those "owning" the accident, thus resolving the issue of symbolic and practical control of the accident. Once the issue of "ownership" was settled, only certain solutions were seen as relevant, which meant that the problems – that is, what the hazards were – were constructed accordingly. The outcome meant, for example, that the scope of medical surveillance became very narrowly defined and excluded potentially important symptoms (for example, reproductive failures and mental health problems) that were central to the exposed individuals. The case also clearly shows that formal risk analyses did not drive decisions as to what should be done; instead these analyses emerged *after* competing definitions of acceptable risks had been winnowed out. Clark concludes that risk assessments in controversial situations are likely to reflect alternatives already chosen and that – similarly to the work within S&TS that was discussed earlier – constructions of risk are *effects* of the outcomes of power struggles rather than arguments used to reach such outcomes.

This finding points to a shared feature of the work of Wynne and Clarke, namely the view that processes of interpreting risks and hazards are *politics-in-the-making*. Emergent constructions of risk are the outcomes of conflicts and power struggles between actors with competing definitions of the risk objects, as well as differential access to resources that can be mobilized to control, manipulate or "own" the risk objects. To understand emerging interpretations of risk and hazards, one thus has to examine the situational rationalities and heterogeneous practices of the various actors, as well as the often conflicting processes by which knowledges, authority and politics are intertwined in attempts to achieve specific outcomes.

Summarizing our approach

The authors in this book utilize a number of insights from the sociological and cultural approaches to risk discussed in this introductory chapter. We have acknowledged our debt to these approaches, but we have also pointed to some of their differences in relation to the studies that we undertake here. In this final section, we will summarize our perspective.

The approach taken in this book focuses on the processes by which risks are interpreted and handled within and among groups of actors in *complex organizations and sociotechnical systems*. We emphasize the importance of understanding the *everyday interactive practices* through which these groups construct safety. From this perspective, risks are collective – but often contested – outcomes that emerge in situated experiences and socially embedded interactions. An important task is thus to analyze the *interpretations and meanings* that actors ascribe to both risk phenomena and the interactions through which these evolve. Our approach thus involves analyzing the dynamic intertwining of knowledge, politics and authority through which interpretations of risks are negotiated, contested and become stabilized at specific sites.

To illustrate the types of analysis that this perspective implies, let us return to the *Estonia* passenger ferry catastrophe introduced in the beginning of this chapter. An analysis along the lines suggested above might mean analyzing the ways in which different groups of actors – for example, crew members, survivors, relatives to the deceased, experts on shipbuilding, journalists, rescue workers – constructed and negotiated conflicting accounts of the series of events that led to the capsizing and its aftermath. Another approach might be to follow the negotiations, conflicts and politics by which official investigative committees in Estonia and Sweden sought to identify the factors behind the catastrophe, distribute responsibility for specific problems or failings and prescribe measures to avoid future accidents. Such a study would involve understanding the interacting practices of powerful actors within organizations such as insurance companies, shipbuilding concerns, major newspapers and trade unions. How did interpretations of the technological risks evolve in interactions and conflicts among the ferry's surviving crew members, members of the investigative committees and other groups – and by what means did the various actors seek to stabilize their views as "facts" of the case?

In our view, the "facts" and the "risks" of this case should clearly be viewed as sociopolitical achievements: they are the outcome of power struggles, contestations and competitions among groups of actors with different values, perspectives and resources. The *Estonia* case is particularly instructive in that – like Wynne's study of events in Cumbria after Chernobyl – it involved a number of "lay experts" that challenged the representations of powerful official investigative committees.

The purpose of this book is to contribute to understanding *risk, uncertainty and safety as heterogeneous phenomena-in-the-making.* The book is organized around three focal areas: interpreting accidents (Part I), defining risks (Part II) and constructing safety (Part III). Each part includes introductory comments that introduce the texts, identify main themes and provide suggestions for future research. The various chapters discuss a number of specific technologies, such as air traffic control, nuclear waste, automobiles, medical technologies and others. Among the groups of actors whose practices and interactions are analyzed are: nurses and doctors in intensive care units, aircraft collision investigators, workers in a printing company, safety engineers in a production and assembly plant, authors of a medical guidebook, laypersons who participated in an Internet site, workers at a plutonium handling facility and experts within auto industries, environmental groups and government agencies. The authors of the various studies draw upon methodologies from ethnography, ethnomethodology, sociology and discourse analysis. There is thus considerable diversity in the cases discussed, but there is also – as we have tried to show in this chapter – a shared concern for the everyday constituting of risks, hazards and safety.

References

Adam, B., Beck, U. and van Loon, J. (eds) (2000) *The Risk Society and Beyond: Critical Issues for Social Theory*, London: Sage.

Adelswärd, V. and Sachs, L. (1998) "Risk Discourse: Recontextualization of Numerical Values in Clinical Practice," *Text* 18: 191–210.

Beck, U. (1992) *Risk Society: Towards a New Modernity*, London: Sage.

—— (1995) *Ecological Politics in an Age of Risk*, Cambridge: Polity Press.

Bellaby, P. (1990) "To Risk or Not to Risk: Uses and Limitations of Mary Douglas on Risk-acceptability for Understanding Health and Safety at Work and Road Accidents," *Sociological Review* 38(3): 465–483.

Boholm, Å. (1996) "Risk Perception and Social Anthropology: Critique of Cultural Theory," *Ethnos* 61(1–2): 64–84.

Bowker, G.C. and Star, S.L. (1999) *Sorting Things Out: Classification and its Consequences*, Cambridge, MA: MIT Press.

Bradbury, J.A. (1989) "The Policy Implications of Differing Concepts of Risk," *Science, Technology & Human Values* 14(4): 380–399.

Caplan, P. (2000) "Introduction: Risk Revisited," in P. Caplan (ed.) *Risk Revisited*, London: Pluto Press.

Clarke, L. (1989) *Acceptable Risk? Making Decisions in a Toxic Environment*, Berkeley: University of California Press.

Clarke, L. and Short, J.F., Jr (1993) "Social Organization and Risks: Some Current Controversies," *Annual Review of Sociology* 19: 375–399.

Cohen, M.J. (ed.) (2000) *Risk in the Modern Age: Social Theory, Science and Environmental Decision-Making*, London: MacMillan Press Ltd.

Douglas, M. (1992) *Risk and Blame: Essays in Cultural Theory*, London: Routledge.

Douglas, M. and Wildavsky, A. (1982) *Risk and Culture*, Berkeley: University of California Press.

Forstorp, P.-A. (1998) "Reporting on the MS *Estonia* Catastrophe: Media-on-media Events and Interprofessional Boundary Work," *Text* 18: 271–299.

Fox, N. (1999) "Postmodern Reflections on 'Risk', 'Hazards' and Life Choices," in D. Lupton (ed.) *Risk and Sociocultural Theory: New Directions and Perspectives*, Cambridge: Cambridge University Press.

Freudenburg, W.R. and Pastor, S.K. (1992) "Public Responses to Technological Risks: Toward a Sociological Perspective," *The Sociological Quarterly* 33: 389–412.

Giddens, A. (1990) *The Consequences of Modernity*, Cambridge: Polity Press.

—— (1991) *Modernity and Self-Identity: Self and Society in the Late Modern Age*, Cambridge: Polity Press.

Green, J. (1997) *Risk and Misfortune: The Social Construction of Accidents*, London: UCL Press.

Hajer, M. (1995) *The Politics of Environmental Discourse: Ecological Modernization and the Policy Process*, Oxford: Oxford University Press.

Halpern, J.J. (1989) "Cognitive Factors Influencing Decision Making in a Highly Reliable Organization," *Industrial Crisis Quarterly* 3: 143–158.

Hannigan, J.A. (1995) *Environmental Sociology: A Social Constructivist Perspective*, London: Routledge.

Haukelid, K. (1999) *Risiko og Sikkerhet: Forståelser og styring*, Oslo: Universitetsforlaget AS.

Heimer, C. (1988) "Social Structure, Psychology, and the Estimation of Risk," *Annual Review of Sociology* 14: 491–519.

Hilgartner, S. (1992) "The Social Construction of Risk Objects: Or, How to Pry Open Networks of Risk," in J.F. Short, Jr and L. Clark (eds) (1992) *Organizations, Uncertainties and Risk*, Oxford: Westview Press.

Jasanoff, S. (1991) "Acceptable Evidence in a Pluralistic Society," in R.D. Hollander and D.G. Mayon (eds) *Acceptable Evidence: Science and Values in Risk Management*, Oxford: Oxford University Press.

—— (1993) "Bridging the Two Cultures of Risk Analysis," *Risk Analysis* 13(2): 123–129.

Jasanoff, S., Markle, G.E., Petersen, J.C. and Pinch, T. (eds) (1995) *Handbook of Science and Technology Studies*, London: Sage Publications.

Jörle, A. and Hellberg, A. (1996) *Katastrofkurs: Estonias väg mot undergång*, Stockholm: Natur och Kultur.

Knorr-Cetina, K. (1999) *Epistemic Cultures: How the Sciences Make Knowledge*, Cambridge, MA: Harvard University Press.

La Porte, T.R. and Consolini, P.M. (1991) "Working in Practice But Not in Theory: Theoretical Challenges of 'High-Reliability Organizations'," *Journal of Public Administration Research and Theory* 1(1): 19–47.

Lash, S., Szerszynski, B. and Wynne, B. (eds) (1996) *Risk, Environment and Modernity: Towards a New Ecology*, London: Sage Publications.

Latour, B. (1987) *Science in Action: How to Follow Scientists and Engineers Through Society*, Cambridge, MA: Harvard University Press.

—— (1999) *Pandora's Hope: Essays on the Reality of Science Studies*, Cambridge, MA: Harvard University Press.

Lidskog, R. and Litmanen, T. (1997) "The Social Shaping of Radwaste Management: The Cases of Sweden and Finland," *Current Sociology* 45(3): 59–79.

Lidskog, R., Sandstedt, E. and Sundqvist, G. (1997) *Samhälle, risk och miljö*, Lund: Studentlitteratur.

Linell, P. and Sarangi, P. (1998) "Discourse Across Professional Boundaries," special issue of *Text*, 18.

Löfstedt, R. and Frewer, L. (eds) (1998) *The Earthscan Reader in Risk and Modern Society*, London: Earthscan Publications Ltd.

Lupton, D. (1999) *Risk*, London and New York: Routledge.

—— (ed.) (2000) *Risk and Sociocultural Theory: New Directions and Perspectives*, New York: Cambridge University Press.

Mol, A.P.J. and Spaargaren, G. (1993) "Environment, Modernity and the Risk-society: The Apocalyptic Horizon of Environmental Reform," *International Sociology* 8(4): 431–459.

Nelkin, D. (ed.) (1979; 2nd edn 1984) *Controversy: the Politics of Technical Decisions*, Beverly Hills: Sage.

—— (1989) "Communicating Technological Risk: The Social Construction of Risk Perception," *Annual Review of Public Health* 10: 95–113.

Perrow, C. (1984; 2nd edn 1999) *Normal Accidents: Living with High-Risk Technologies*, Princeton: Princeton University Press.

Rayner, S. (1986) "Management of Radiation Hazards in Hospitals: Plural Rationalities in a Single Institution," *Social Studies of Science* 16: 573–591.

—— (1992) "Cultural Theory and Risk Analysis," in S. Krimsky and D. Golding (eds) *Social Theories of Risk*, Westport, CT: Praeger.

Reason, J. (1990) *Human Error*, Cambridge: Cambridge University Press.

Renn, O. (1992) "Concepts of Risk: A Classification," in S. Krimsky and D. Golding (eds) *Social Theories of Risk*, Westport, CT: Praeger.

Roberts, K. (1990) "Managing High Reliability Organizations," *California Management Review* 32(4): 101–113.

Rochlin, G.I. (1991) "Iran Air Flight 655: Complex, Large-Scale Military Systems and the Failure of Control," in R. Mayntz and T.R. La Porte (eds) *Responding to Large Technical Systems: Control or Anticipation*, Amsterdam: Kluwer.

Rosa, E.A. (1998) "Metatheoretical Foundations for Post-normal Risk," *Journal of Risk Research* 1: 15–44.

Schwarz, M. and Thompson, M. (1990) *Divided We Stand: Redefining Politics, Technology and Social Choice*, New York: Harvester Wheatsheaf.

Shrivastava, P. (1987) *Bhopal: Anatomy of a Crisis*, Cambridge, MA: Ballinger Pub. Co.

Sjöberg, L. (1997) "Explaining Risk Perception: An Empirical Evaluation of Cultural Theory," *Risk Decision and Policy* 2: 113–130.

Sundqvist, G. (2002) *The Bedrock of Opinion: Science, Technology and Society in the Siting of High-Level Nuclear Waste*, Dordrecht: Kluwer Academic Publishers.

The joint accident investigation commission of Estonia, Finland and Sweden (1997) *Final Report on Capsizing on 28 September 1994 on Baltic Sea of the Ro-Ro Passenger Vessel Mv. Estonia.*

Turner, B.A. (1978; 2nd edition with N.F. Pidgeon 1997) *Man-made Disasters*, Oxford: Butterworth-Heinemann.

Vaughan, D. (1996) *The Challenger Launch Decision: Risky Technology, Culture, and Deviance at NASA*, Chicago: University of Chicago Press.

—— (1999) "The Dark Side of Organizations: Mistake, Misconduct, and Disaster," *Annual Review of Sociology* 25: 271–305.

Vera-Sanso, P. (2000) "Risk-talk: The Politics of Risk and its Representation," in P. Caplan (ed.) *Risk Revisited*, London: Pluto Press.

Weick, K.E. (1987) Organizational Culture as a Source of High Reliability," *California Management Review* XXIX(2): 112–127.

Wilkinson, I. (2001) "Social Theories of Risk Perception: At Once Indispensable and Insufficient," *Current Sociology* 49(1): 1–22.

Wynne, B. (1989a) "Frameworks of Rationality in Risk Management: Towards the Testing of Naïve Sociology," in J. Brown (ed.) *Environmental Threats: Perception, Analysis and Management*, London: Belhaven Press.

—— (1989b) "Sheepfarming after Chernobyl: A Case Study in Communicating Scientific Information," *Environment*, 31(2): 11–39.

—— (1992) "Risk and Social Learning: Reification to Engagement," in S. Krimsky and D. Golding (eds) *Social Theories of Risk*, Westport, CT: Praeger.

Part I

Interpreting accidents

Introductory comments

The chapters in Part I center on problematic understandings of the accidental. Judith Green discusses how technologies and knowledge systems have constructed "the accident" as a particular category of misfortune in late modern society. Jörg Potthast and Sidney W.A. Dekker focus in different ways on the discursive constructions of what caused specific accidents in sociotechnical systems (specifically air transport systems). The studies point to the various strategies by which actors attempt to make sense of injuries and catastrophic events, continually striving to control "the accidental."

In an important recent book, Green (1997) traces the development of European thought in relation to the accident. She identifies three dominant discourses that could be crudely summarized as "fate," "determinism" and "risk." Before the second half of the seventeenth century, events in the course of one's life were viewed as part of one's personal destiny. Accidents as something that "just happened" had no place in life since all events had to fit into a pattern. By the end of the seventeenth century, this view of fate and destiny had been replaced by a modernist discourse of determinism. Accidents were now a residual category of misfortune that marked the limits of a rational universe. Given the vagaries of "nature," accidents were inevitable and thus to be expected from time to time – yet not worthy of serious investigation. This consensus was fractured in the second half of the twentieth century. In today's "risk society," a discourse of risk management and prevention has emerged, marking the transfer of "the accidental" from the margin to the center of concern. If misfortunes such as accidents occur as a result of knowable and calculable risks, they can be transformed into potentially preventable events. All accidents thus become comprehensible in terms of "risk factors," and "the accidental" becomes something to avoid.

These themes are further developed in Green's chapter in this part. She focuses on the science of epidemiology as the dominant framework for generating knowledge about accidents since the mid-twentieth century. Epidemiology is viewed as a technology for understanding accidents as the outcome of mismanaged risks. Accidents have been reconfigured as not only potentially preventable but as events that *ought* to be prevented. The technologies of epidemiology and the artifacts that they make possible (such as safety devices for reducing risks) have, she argues, given rise to an individualization of accident management. Everyone is responsible for making rational decisions about avoiding and managing risks, using the everyday risk technologies that are available. Individuals are accountable for their personal misfortunes. There is a gap, however, between the myth of preventability and the occurrence of what should have been prevented. The risks are separated from the uncertain and fluid life-world in which they are experienced; they are fragmented and reified as specific "risks" to be calculated, catalogued and monitored. The ways in which risks and the accidental are classified and handled are thus seen as processes of both construction and control. This focus links Green's analysis to work on the disciplinary nature of knowledge and expertise in the Foucaldian tradition (see, for example, Burchell *et al.* 1991; Lupton 1999). Yet Green does not only discuss the proliferation of safety technologies but also the resistances they engender, based on the individual's chaotic experience of everyday life.

Until the second half of the twentieth century, there was little difference between the "expert" view of accident causation and that attributed to the public, which Green points out. Since then, professionals have tried to move away from a "popular" understanding of accidents toward a uniquely "professional" understanding based on what is viewed as a rational approach to risks and their management. The "irrationality" of the popular understandings was then – as was discussed in the introductory chapter to this volume – to be studied in psychological and psychometric work. Jörg Potthast gives an interesting perspective on these emerging tensions between expert and lay opinions in his study of different interpretations of the causes of the Swissair crash near Halifax, Canada, in 1998. He analyzes a large number of contributions to a discussion site on the Internet in the aftermath of the crash. Potthast's analysis shows that the boundaries between expert and lay opinions are blurry at best, despite the attempts of some contributors to invoke an authoritative voice.

Potthast's analysis shows the ways in which reconstructions of accidents by both expert and lay actors are often characterized by a high degree of uncertainty. Actors with differing competences and orientations struggle to reconstruct and understand a sequence of events in which the outcome may be clear but the process leading to this outcome is largely unknown. As various explanations of the accident are contested or reaffirmed, fragments of a negotiated understanding emerge among the contributors to the website. Potthast views the sense-making that is involved in these negotiations as reflecting three distinct "narratives of trust" that show how the actors work to reconstruct their trust in the artifacts, the operators and other experts and the safe operation of the system.

The chapter by Sidney W.A. Dekker focuses on another theme from Green's analysis, that is, the tendency for expert accounts to be individualized and reified. Dekker discusses the accounts given by official investigating commissions in the wake of another, less publicized airline accident, namely a 1995 crash in Cali, Columbia, that resulted in the deaths of all on board the flight. Dekker's story provides a detailed and incisive account of the tendency of investigators to confuse their own reality with the crew's situation at the time of the accident. In these official accounts, the crew's actions are removed from the context in which they are embedded and are ascribed with meaning in relation to after-the-fact-worlds. Such accounts exhibit all the errors of hindsight: through the matching of available and isolated fragments of behavior, an artificial story that is divorced from the situated reality of the actors at the time is successively constructed. Dekker convincingly argues for a different approach to accident investigations which would take into account the explicitly situated character of cognition and decision making. Dekker's analysis thus aligns with the work of Karl Weick, Gene Rochlin and others whose studies of so-called High Reliability Organizations explore the contextual rationality and sense making of actors in situations of stress and uncertainty (Rochlin 1991; Weick 1993; see also the introductory chapter to this volume).

References

Burchell, G., Gordon, C. and Miller, P. (eds) (1991) *The Foucault Effect: Studies in Governmentality*, London: Harvester Wheatsheaf.

Green, J. (1997) *Risk and Misfortune: The Social Construction of Accidents*, London: UCL Press.

Lupton, D. (ed.) (1999) *Risk and Sociocultural Theory: New Directions and Perspectives*, Cambridge: Cambridge University Press.

Rochlin, G.I. (1991) "Iran Air Flight 655: Complex, Large-scale Military Systems and the Failure of Control," in T.R. La Porte (ed.) *Social Responses to Large Technical Systems: Control or Anticipation*, Amsterdam: Kluwer.

Weick, K.E. (1993) "The Collapse of Sensemaking in Organizations: The Mann Gulch Disaster," *Administrative Science Quarterly*, **38**: 628–652.

1 The ultimate challenge for risk technologies

Controlling the accidental

Judith Green

Introduction

This chapter addresses technologies of risk in one arena, that of accidental injury. "Technologies of risk" refers here to artifacts and knowledge systems that are utilized for the management and reduction of risk. Accidents present perhaps the ultimate challenge for risk technologies – how to control apparently unpredictable events. However, risk technologies do not merely act upon a natural category of events called "accidents," they also simultaneously shape the classification of certain events as "accidents" and shape our understanding of and reactions to those events. This chapter explores this simultaneous process of control and construction through an analysis of how two interrelated sets of technologies have constructed "the accident" as a particular category of misfortune in late modern society.

The first technologies of interest are those mundane and unremarkable artifacts that are designed to reduce the risk of an accident happening or reduce the risk of injury if an accident should occur. These include such innovations as child-resistant medicine containers, bicycle helmets and airbags in cars. These particular examples are now commonplace and accepted as sensible ways of increasing safety – no rational adult would resist their use. However, the introduction of such risk technologies is not inevitable and straightforward, and the example referred to later in this chapter (namely "soft hip protectors" for elderly citizens) is of a more recent and less widely accepted artifact.

Second, are those technologies of knowledge that make possible risk reducing artifacts. From the second half of the twentieth century, legitimate knowledge about accidental injury has been primarily derived from statistical analysis of the distribution of accidents and their risk factors within populations. As accidental injury has become "medicalized," epidemiology, the study of diseases and their distribution throughout populations, has become a dominant framework for generating knowledge about accidents. Given that the term "accident," as has already been indicated, is taken here to describe a socially constituted category of events, this chapter will start not by trying to define the term, but by delineating the historical space within which accidents have happened. The question is not so much, "what is an accidental injury?" but rather "when did it become possible to speak of accidental injuries?" After outlining the historical emergence of accidents, this chapter describes the development of epidemiology as a technology for understanding accidents as the outcome of mismanaged risks. Two implications of an epidemiological understanding are then identified: the individualization of risk management and the separation and reification of risks. Finally, this chapter explores the possibilities for resisting risk technologies.

Historical context

As a category of misfortune, "accidents" have a relatively brief history (Green 1997a). In Europe, they are essentially products of the seventeenth century, when emerging rationalist and probabilistic ways of thinking (Hacking 1975; Porter 1986) created a space for those misfortunes which were, at least at the local level, inexplicable. Before then it could be argued that Western cosmologies had no need for a category of accidental events. Fate, destiny and a belief in predetermination meant all happenings could have meaning. With religion and superstition declining in legitimacy as answers to the inevitable questions of "why me, and why now?" when misfortune strikes, the accident became first a possible, then a necessary, category. By the beginning of the twentieth century, a belief that some events happen "by accident," that they have no purpose and cannot be predicted at the local level, had become a defining belief of modernity.

That accidents came to signal a uniquely "modern" system of beliefs is evident in the writings of both European anthropologists, anxious to distance the superstitious beliefs of "primitives" from their own, and in those of child psychologists newly interested in the development of "adult" mentality. Anthropologists such as Evans-Pritchard (1937) pointed to the cosmologies of small-scale societies such as the Azande in central Africa. For the Azande, argued Evans-Pritchard, all misfortune had meaning and could be traced to witchcraft or the breaking of taboos. The specific cause of any misfortune (that is, why it happened to this person, at this time) could be divined. For Evans-Pritchard and others looking at the apparently irrational ideas of non-Western peoples, the belief that the local distribution of misfortune was random, unpredictable and inexplicable was what distinguished the modern rational mind from that of the "primitive" who persisted with such supernatural explanations.

Similarly, for child psychologists such as Piaget in the 1930s, children's failure to understand chance and coincidence were evidence of their immature minds (Piaget 1932). Moral maturity brought with it the ability to conceptualize the accident as a blameless event. In the early twentieth century, then, a belief that accidents "just happen," and that there is therefore no profit in examining why they happen to particular people at particular times, was a legitimate and rational belief. The accident was merely a coincidence in space and time with neither human nor divine motivation. Only those yet to develop a modernist rationality (primitives, children) would see an accident as anything other than a random misfortune.

More recently the place of accidents as a category of motiveless and unpredictable misfortunes has shifted. From a marginal category of events not worthy of serious study, they have moved center stage. Within the risk society, accidents are pivotal. First, management of the accidental poses the ultimate challenge for risk technologies: to predict the unpredictable and then control the uncontrollable. Second, that accidents do happen demonstrates the failings of risk management to date and thus justifies even greater surveillance of risks. Despite ever more sophisticated safety technologies, trains occasionally crash and children frequently fall from bicycles. Major tragedies and minor everyday mishaps are subjected to formal or informal enquiries to establish "what went wrong" and to apportion blame if appropriate. Such enquiries apparently erase the notion of the accidental, for in retrospect all misfortunes are rendered comprehensible in terms of the risk factors that produced them. They were therefore not accidents at all, but rather failures of risk management. Like the Azande in Evans-Pritchard's (1937) account, we can no longer, it seems, countenance the idea of a meaningless event, a misfortune "caused" merely by bad luck or coincidence (Douglas and Wildawsky 1982). However, risk technologies of course have not erased the accidental at all. It survives and even flourishes but has been shaped by the very technologies that seek to manage it. To illustrate the ways in which risk technologies, and epidemiological technologies in particular, contribute to particular understandings of the accident, this chapter will now outline the growth of epidemiological technologies of knowledge.

The growth of epidemiology

As a professional activity, accident prevention has an even briefer history than that of the accident. From a vantage point of the early twenty-first century, it is perhaps self-evident and unquestionable that accidents should be "prevented" if possible. However, the possibility of organized prevention activities did not become thinkable until accidents themselves were framed as patterned by their underlying causes, which could then be amenable to potential manipulation. Also, until the second half of the twentieth century, there was little difference between the "expert" view of accident causation and that attributed to the public. Both held accidents to be distributed randomly across space and time and thus not a proper object of scientific scrutiny in order to determine causes – unless they were associated with dramatic death rates in the hazardous industries of mining and transport. The concern of health professionals was with accident services and the rehabilitation of victims, not identifying the causes of accidents and potential ways of reducing their incidence. In the reports and academic journals of early twentieth century public health, there is little that could be recognized as an organized or professional approach to accident prevention (Green 1999).

A turning point was John E. Gordon's (1964) paper on the epidemiology of accidents, first published in 1949, but not widely cited until over a decade later. Gordon used the analogy of trench foot in war to argue that medicine ought to be properly concerned with the prevention of accidental injury. Wartime experiences suggested that the incidence of trench foot, originally thought to be caused by cold and therefore not amenable to medical research and intervention, was affected by foot hygiene which could be improved through careful investigation as to which men became diseased and what factors contributed to their affliction. In peacetime, Gordon argued, accidents posed a similar challenge. Although they might appear as inevitable features of the natural world (much like cold), a careful analysis of the interactions between hosts, agents and vectors would reveal regular patterns. Like many endemic diseases, certain accident rates (such as those for home accidents) appeared remarkably regular from year to year. Others (such as road traffic accidents) mimicked the seasonal incidence rates of other diseases. "Accidents," argued Gordon, "evidently follow as distinctive movements in time as do diseases" (Gordon 1964: 20). As such, they could be brought within the realm of public health medicine, with their epidemiology mapped and interventions devised to reduce their incidence.

From the 1950s onwards, professionals began to speak explicitly of moving away from a "popular" understanding of accidents as unmotivated coincidences of time and space and towards a uniquely professional understanding that was based on a rational approach to risks and their management. This emergent "rational" approach involved collating facts about the incidence, types and settings of accidental injuries and subjecting these data to increasingly sophisticated statistical analysis in order better to understand the predictors of injury and their interrelationships. The range of variables related to general or specific accidental injury risks is now vast, including demographic factors; psychological, attitudinal and behavioral traits; structural factors (unemployment rates, measures of social inequality); environmental contexts and legal codes. As many of these factors are amenable to intervention through public education about risk, environmental change or legislation, public health could gradually colonize accidental injury as a proper object of concern and speak with a self-consciously "professional voice" about prevention. This professional voice situated itself as explicitly opposed to fatalistic "lay" beliefs about accidents, which supposedly belonged to an earlier regime in which accidents "just happened" and were unpredictable and therefore unpreventable. The professionalization of accident prevention generated a class of experts in the production of accident knowledge and the development of risk technologies.

In these new "expert" accounts, the public was portrayed as clinging to an anachronistic and irrational view of the "accidental" accident. For Haddon, who was one of the founders of public health research into accidents, the key to success in this emerging field was ridding science of what he called the "common sense approach." He argued that "the folklore of accidents is perhaps the last folklore subscribed to by rational men" (Haddon *et al.* 1964: 6). Such views about the irrational fatalism of the public have persisted throughout the twentieth century and through to the present day. One writer noted that: "[a]ccidents are not totally random events striking innocent victims like bolts from the blue, although they are often described in this way" (Stone 1991). In literature aimed at public health professionals and at potential victims and their caretakers, the message that it was "vital to counter the view that accidents are random events due to bad luck" (Henwood 1992) became ubiquitous.

The identification of accidents and their prevention as a legitimate arena for public health interest brought with it a mushrooming of research into the "patternings" of accidental injuries. The epidemiology of accidents became ever more sophisticated during the 1970s and 1980s, with studies identifying the risks of injury in all fields of life and gradually identifying the interrelationships among those risks. Accidental injury is demonstrably non-random in that its epidemiology can be mapped. With this mapping comes the knowledge necessary for prevention. Ultimately, therefore, accidents can be prevented. This notion, that is, that increasing knowledge of risks and their correlates implied that they could be managed and that accidents could be prevented, has of course been enthusiastically adopted by both professional associations and policy makers throughout Europe. For example, in 1993 Britain's Department of Health claimed that "Most accidents are preventable" (DOH 1993: 9), and in Sweden there is now a national "Zero Vision" for road traffic accidents, based on the notion that such accidents can, at least in theory, be completely abolished.

Epidemiology and the non-accidental accident

Thus within a few decades accidents had been reconfigured as not only potentially preventable but as events that *ought* to be prevented. The occurrence of an accident was no longer a random misfortune, but rather an indication of risks being mismanaged. The accident is therefore not an accident – there is nothing left which could be unexpected and unmotivated. One example, taken from the news pages of the journal *Injury Prevention* in 1999, illustrates perhaps the narrowing range of events which can still be viewed as "purely tragic misfortunes":

Update on an airbag fatality

A man whose infant son was killed by an airbag in Ohio, USA was fined $500 and sentenced to two 12 hour days in jail for failing to turn off the device. The father will serve his time on the boy's birthday and on the anniversary of the crash, the municipal court judge determined. This is believed to be the first case of its kind in North America. The 29 year old father pleaded no contest on charges of vehicular homicide and running a red light in the death of his 2 month old son. They were in a pick up truck with a switch off device for the passenger side airbag (and prominent warning signs) in the event a child was riding in that seat. The infant was in a rear-facing car seat.

(*Injury Prevention* 1999: 5: 123)

This news item is published in the journal without editorial comment, so there is scope for various potential readings. The item could be read as a comment on the vagaries of the judicial system of the US or on the carelessness of drivers or as suggesting that, morally, all is as it should be and blame has been properly apportioned. What is evident through all these readings, however, is that this extreme misfortune has clearly been defined as "other" than accidental. It has been categorized as a homicide and its perpetrator punished accordingly. What is also evident is the expectation of skills from responsible adults, who must not only be vigilant of the risks of the road (e.g. red lights, other road users) but also be responsible and competent in the use of safety technologies. In this case competence relies on literacy and comprehension skills to utilize one of the many safety notices we are faced with every day, as well as the technical skills needed to disable the air-bag.

So one could argue that the accident has indeed disappeared, a relic of a particular moment in history between the seventeenth and twentieth centuries and a marginal category of misfortune necessary only until the hyper-rationalization of high modernity "tamed chance" (Hacking 1990) itself and made the random predictable. In late modernity, there can be no accidents, only risks mismanaged or inadequately mapped. The logical outcome for public health professionals has been attempts to erase the very word "accident" from English language medical journals. Both the specialist journal in the field, *Injury Prevention*, and the generalist journal for the profession in the UK, the *British Medical Journal*, have now attempted to ban the word "accident" from their pages (Bijur 1995; Davis and Pless 2001). Writers must instead use terms such as "injury" which are less tainted with notions of unpredictability. Such moves are perhaps illustrative of what Castel has called "the grandiose technocratic rationalizing dream of absolute control of the accidental" (Castel 1991: 289). The ultimate outcome of increased surveillance of risks and the epidemiological approach to prevention is the eradication of chance events.

If a belief that some misfortunes "just happened" was a defining belief of early twentieth century modernism, the belief that all misfortune can (at least in theory) be managed is perhaps the defining belief of the risk society. That Castel calls this a "dream," however, points to the inherent tensions in attempts to eradicate accidents. The very technologies that are designed to map, control and ultimately erase accidents in themselves of course simultaneously generate a proliferation of accident discourses. Epidemiological technologies attempt erasure (through rendering all accidents non-accidental) but at the same time produce accidents through endless categorization and investigation. Yet this production is not infinitely variable: risk technologies produce particular kinds of accidents, and make particular responses to them both possible and legitimate. Two features of the epidemiological accident are its individualization and the reification of its risks, which will now be discussed.

The individualization of accidents

The technologies of epidemiology and the products they make possible (namely safety devices for reducing risks) are at one level irresistible because they are self-evidently rational. To manage risks is to construct oneself as a rational, competent human being who is capable of keeping oneself and any dependants safe. To refuse to wear a helmet when riding a bicycle is to present oneself as an irrational risk taker; not to choose the packaging with child-resistant opening is to present oneself as an irresponsible parent. The utilization of everyday risk technologies thus has a moral as well as a technical dimension in that it demonstrates not only proper prudence in the face of a hazardous world but also a commitment to the whole risk management enterprise. To wear a helmet or install child safety equipment symbolizes a faith in the possibility of controlling the accidental, and says "I did all I could" in the event that an accident does occur.

This construction of responsibility for safety as a marker of maturity has been noted even in young people. In studies of children's accounts of accidents and how to prevent them (Green 1997b) even the youngest respondents, namely those aged between seven and twelve years old, assumed that safety was their own responsibility and not that of parents or other caretakers. In telling stories about accidents that have happened to them, children talked about themselves as having primary responsibility for assessing risks and accepting responsibility when risk assessment went wrong. Only when discussing younger siblings or their younger selves did they hold others as having responsibility for safety. Presenting yourself as a mature individual involved presenting yourself as responsible for managing the risks you faced.

The understanding and management of misfortune in the late twentieth century is predicated on our ability to monitor and manage risks in an ever-widening range of arenas. The production of legitimate knowledge about accidents has been professionalized, but the nature of this "expertise" is that it must be disseminated throughout the public. Experts may be central to the development of epidemiological technologies, but it is individuals who must take responsibility for utilizing risk-reducing technology. In short, the responsibility for this risk assessment has become individualized. Epidemiology delineates the field of potential risks and professionals inform the public of how to reduce those risks, but the task of selecting (from a potentially limitless field), assessing and managing those risks lies largely with the individual. We are all engaged in a seemingly irresistible strategy of constant risk management.

An illustration of this individualization and its implications is a book prepared for the public by the British Medical Association (BMA), the *BMA Guide to Living with Risk* (1990). The BMA suggested that one purpose of their guide was to "put risk in perspective, and to put numbers on a selection of risks as far as possible" (BMA 1990: xvii). In earlier years faith in such a rational analysis of numbers would perhaps have been reassuring, a matter of extending our understanding of the physical world to provide consensus expert advice for the public. Further knowledge would correct misconceptions and the quantification of that knowledge would lead to reduced anxiety. The BMA guide is, however, profoundly unsettling. "It is not," the authors point out, "possible to make choices for people" (BMA 1990: xviii). People must instead make their own choices but from an ever-extending array of possible risks which must be quantified, understood and balanced. Rather than reducing anxiety, knowledge of risks increases it. There are, note the authors, "few things that are certain in this uncertain and complex world" (BMA 1990: xv), and the range of risks we face is vast:

> One might ask, how might it affect me if a nuclear power station was built nearby, rather than a coal-fired one? If I collide with that car over there, would I be safer in it or my own? . . . How often does someone check the brakes of the train I am travelling in tomorrow?
>
> (BMA 1990: xv)

Transport, food, leisure activities, work and medical care all involve sets of risks which have been studied and which the BMA here report to a lay audience in order that they can make rational decisions about managing risks. In a world in which expert production of knowledge has been separated from responsibility for management, risk assessment is explicitly an individualized activity, in which we must all constantly engage.

The reification of risk

Accidents happen not to epidemiological populations, but to individuals who must account for their personal misfortune. One problem, of course, is that epidemiological knowledge refers to populations, not to the individuals who suffer from mismanaged risks. Castel's "grandiose rationalizing dream" remains a dream because of this tension between the sophistication of knowledge about risks in populations and the paucity of understanding that can be applied to the individual level. There is a gap between the myth of preventability (in a population where all are aware of their personal responsibility to manage risks) and the occurrence of what ought to have been prevented. An ironic illustration of this tension comes from the new year's edition of a national British newspaper (*Sunday Times*, January 1991) which, under the heading the "year of glorious follies," noted that the Department of Transport had worked out the risk of having a road accident. It went on: "The formula you have to apply is this"

$$A_c = 0.00633 \exp\{s + g\}$$
$$(1 = 1.6p_d)$$
$$(p_b + 0.65p_r + 0.88p_m)$$
$$M^{0.279}$$
$$\exp\{b_1/Ag + b_2/(X + 2.6)\}$$

Such mathematical precision is testament to the advanced sophistication of epidemiological technology, but it is clearly of limited use for informing the everyday business of attempting to cross the road or drive a car without injury. The "dream" remains just that in the face the messy, dangerous and uncertain world of the street, where one has to balance the needs for transport and perhaps pleasure with the risks from traffic, other people and the environment. A crisp figure, for those with access to the knowledge and mathematical skills to calculate it, constitutes the exact risk of having a road traffic accident, but its ironic use as a front cover illustration of "follies" suggests the uselessness of such knowledge for individual decision making.

So if the first effect of epidemiological constructions is that risk management is individualized, the second effect is that dangers are separated from the uncertain and fluid life-world in which they are experienced, and fragmented and reified as specific "risks" to be calculated, catalogued and monitored. Within the BMA *Guide to Living with Risk* (BMA 1990), one can look up the risk of death from flying in different kinds of aircraft, the risk of serious injury from playing different sports or the risk of adverse drug reactions from different kinds of drugs. Once reified as individual risks rather than the diffuse hazards one might face throughout the normal business of everyday life, behaviors and events can then be subject to studies of potential interventions. In public health medicine, the most prized knowledge is that from the randomized controlled trial (RCT), an experiment in which interventions are tested in conditions that "control for" the other features of everyday life that would potentially interfere with the findings. Educational interventions or artifacts designed to reduce risk can be tested to see if they do indeed reduce the number of accidental injuries. In accident research, the body of evidence from experimental studies is growing, and there have been calls for increasing the number of interventions that are "tested" in this way for their potential in reducing injuries.

One example of an experiment on injury research and the fate of it in practice will perhaps illustrate both the reification of accident risks in epidemiology and the potential for resistance. The example is that of one specific risk technology, a safety device known as the "soft hip protector pad." This device is designed to be worn inside specially designed underwear. It is intended for older people in residential care to reduce the risk of fracturing the hip if a fall occurs. An RCT from Denmark (Lauritzen *et al.* 1993) found a 53 percent reduction in hip fractures in the group of nursing home residents who were given the hip protectors. Of those in this group who did suffer hip fractures, none was wearing hip protectors at the time. This finding was referred to in a bulletin aimed at health policy makers and included a recommendation that the evidence for its effectiveness was convincing enough for the intervention to be recommended. There was also a comment that the acceptability of such devices, especially in community settings, was not clear (NIH 1996: 6). In turn, this finding on the effectiveness of soft hip protector pads as a risk reducing technology is then developed as "good practice" for authorities planning evidence-based interventions to reduce hip fractures within their residential care homes.

From the epidemiological perspective, the management of risks in this case is perhaps straightforward. Fractures cause considerable disability and premature mortality in elderly people, and the effects of an accident can clearly be minimized by advocating the utilization of such proven safety technologies. Technologies of knowledge, including both the epidemiological surveys of injuries in elderly people and RCTs designed to test the effectiveness of interventions, shape particular events (such as hip fractures in residential homes) as preventable and therefore (eventually) non-accidental. Like cycle helmets and child-resistant containers, the soft hip protector pad ought to be irresistible as a safety intervention.

Technology and resistance

However, embedding technology in practice is more troublesome, and it is perhaps only once such artifacts have become taken for granted that they appear as self-evidently rational choices. In Michael's (1997) phrase, new technologies must be "inoculated against ridicule" if they are not to appear absurd. Michael is discussing innovative, everyday technologies like those that are advertised on the back pages of magazines, such as nose clippers or specially shaped pillows. For these artifacts to be presented as necessary aids to daily living rather than absurdities, he argues, "the problem" that they address (such as unwanted nasal hair or the discomfort of reading in bed) must be successfully framed as a deterrent to a good quality of life. Although the soft hip protector pad addresses an already established legitimate problem (hip fractures in elderly citizens), the solution proposed still risks being ridiculed. This technology may "work" at a technical level, but it also of course carries other symbolic meanings as well as those about safety. These include associations of lack of dignity and independence. The reification of one risk (hip fracture) is possible within an epidemiological framework of knowledge but may not reflect the life-world of elderly citizens or those who

care for them. There is, therefore, scope for resisting epidemiological technologies. Some evidence of this resistance is found in a study of Accident
Alliances (Green 2000), which examined how professionals involved in accident prevention work viewed evidence. First, the study found that workers
from a range of professions expressed routine cynicism about the role of the
very technologies which produce the supposedly most valued knowledge
about accident risks. Most health promotion workers and health and safety
officers were dismissive of the worth of routine accident data. The heavy
sarcasm of one environmental health officer, who is commenting on the
routine accident analyses he receives from the government, was typical of
responses to studies and reports that merely found what "everybody knew"
already: "Here – it says '60% of major injuries in offices are slips and trips!'
Well, knock me down with a feather – I could never have worked that out
on my own!" (Green 2000: 464).

Second, most workers who participated in the study, despite being in
involved in public sector partnerships to implement accident prevention
programs, were explicitly skeptical of the value of the RCT for informing
real life-world decisions. As one health promotion worker put it: "Evidence
based medicine is a lot different to community development . . . I mean the
practicalities of these things are never published" (Green 2000: 465).

For many of those included in the study, the example of soft hip protectors
was indeed an archetypal illustration of the limits to epidemiological technology. It had become a kind of moral tale that was used rhetorically to under-
line the distance of "scientists" from the "real world." Health promotion
workers resisted the ways in which knowledge based on experiments and
epidemiology tended to reify the risk of hip fracture, stripping it from the context of everyday life and from values such as dignity and self-determination.
As one health promotion worker put it: "The Health Authority have firm policies on everything being evidence based, using these big reviews . . . I've got
to say I heartily disagree with that – they recommended wearing padded
knickers!" (Green 2000: 471).

At one level the technologies which construct accidents as knowable and
preventable (such as epidemiological statistical techniques or devices such as
soft hip protectors) are those which made possible the professional roles of
the workers included in this study. Health promotion officers, health and
safety officials and environmental health officers are creations of the epidemiological regime of risk, in which risk factors for accidents can be identified
and interventions designed. However, the gap between on the one hand the
population level at which knowledge about accidental injury is produced,
and on the other hand the individual level where it is experienced, is potentially a site of resistance. Workers construct themselves and their professional
identities as having no need for the "experts" of the risk society, namely the
epidemiologists who produce the evidence base for accident prevention. As
individuals, we largely experience safety technologies as irresistible. Who, as
a rational mature adult, could refuse to wear a car seat belt or a cycle helmet,

or refuse to fit guards around the fire to protect their children? It is only when a new technology is being introduced that the less-than-inevitable process of embedding risk technologies into everyday life becomes visible.

Conclusion

Epidemiological technologies reconstruct many misfortunes as "accidental injuries" that can be understood, managed and ultimately eradicated through increased vigilance. However, there is no straightforward way in which these technologies constitute events in the real world. "Accidents" may have disappeared, reconstituted by epidemiology as reified population risks which can be controlled, but the life-world is still uncertain, messy and liable to throw up misfortunes which have to be dealt with as individual events. Epidemiology produces a set of risks for fragmented individuals who share a proportion of a population risk.

Subjectively, though, we still experience badly managed risk as individual misfortune and have often no way of accounting for it. Hence the irresistibility of safety technology at one level. To engage in practical accident prevention through wearing safety helmets or utilizing a car air-bag correctly is not only to demonstrate our faith in the possibilities of risk management, but also to show we have done all we can – that we can be absolved from blame.

However, at the same time, the proliferation of safety technology generates its own resistance. In attempting to "control" the accidental, risk technologies recreate "accidents" on at least three levels. First, to justify the exercise of expertise in both the production of knowledge and the dissemination of artifacts, the supposed irrationality of the public must be constantly reiterated. Thus "accidents" are constantly being named and recreated, if only as a despised category of misfortune utilized by a hypothetical "public." Second, absolute control of the accidental remains, in practice at least, a dream. "Accidents" will still happen. Such events are perhaps even necessary to justify ever-increasing efforts at surveillance and management. Third, risk technologies make particular strategies of management and response more legitimate than others. The most legitimate are those of individualized responsibility for discrete risks. In the gap between the sophistication of population-based knowledge (as seen in the formula for a road traffic accident or the data on hip fractures in the elderly) and the individual's chaotic experience of the everyday world (such as managing transport needs in a busy city or maintaining a valued sense of self in a residential home) lies the potential for reframing technologies as absurdities ("padded knickers") rather than rational responses.

Technologies such as epidemiological surveys and safety devices serve to reduce our risks of accidental injury. They have been developed to map the risks from injury, identify causal factors and intervene to remove or ameliorate those factors. However, such technologies also mediate our understanding of risk and safety by producing categories of misfortune called "accidents" and delineating legitimate responses to them. Such responses shape individuals as ultimately responsible for their own safety: they must assess risks, select those to attend to, and manage their environments adequately to reduce them. In addition, legitimate responses create discrete "risks" to be managed in isolation. However, risk technologies have not – at least not yet – erased the accidental. Instead they have reconstructed accidents as pivotal events of the risk society, providing both the ultimate challenge and the ultimate reminder of the impossibility of erasing misfortune.

Acknowledgments

I would like to thank participants at the workshop for a very helpful discussion on an earlier version of this chapter, and Simon Carter for his comments on an earlier draft.

References

Bijur, P. (1995) "What's in a Name? Comments on the Use of the Terms 'Accident' and 'Injury'," *Injury Prevention* 1: 9.

BMA (British Medical Association) (1990) *The BMA Guide to Living with Risk*, Harmondsworth: Penguin.

Castel, R. (1991) "From Dangerousness to Risk," in G. Burchell, C. Gordon and P. Miller (eds) *The Foucault Effect: Studies in Governmentality*, London: Harvester Wheatsheaf.

Davis, R.M. and Pless, B. (2001) "BMJ Bans 'Accidents'," *British Medical Journal* 322: 1320–1321.

DOH (Department of Health) (1993) *The Health of the Nation: Key Area Handbook – Accidents*, Leeds: Department of Health.

Douglas, M. and Wildavsky, A. (1982) *Risk and Culture*, Berkeley: University of California Press.

Evans-Pritchard, E.E. (1937) *Witchcraft, Oracles and Magic Among the Azande*, Oxford: Clarendon Press.

Gordon, J. (1964) "The Epidemiology of Accidents," in W. Haddon *et al.* (eds) *Accident Research: Methods and Approaches*, New York: Harper and Row.

Green, J. (1997a) *Risk and Misfortune: The Social Construction of Accidents*, London: UCL Press.

—— (1997b) "Risk and the Construction of Social Identity: Children's Talk about Accidents," *Sociology of Health and Illness* 19: 457–479.

—— (1999) "From Accidents to Risk: Public Health and Preventable Injury," *Health, Risk and Society* 1: 25–39.

—— (2000) "Evidence, Epistemology and Experience," *Sociology of Health and Illness* 22: 453–476.

Hacking, I. (1975) *The Emergence of Probability*; Cambridge: Cambridge University Press.

—— (1990) *The Taming of Chance*, Cambridge: Cambridge University Press.

Haddon, W., Suchman, E. and Klein, D. (1964) *Accident Research: Methods and Approaches*, New York: Harper and Row.

Henwood, M. (1992) *Accident Prevention and Public Health: A Study of the Annual Reports of Directors of Public Health*, Birmingham: RoSPA.

Lauritzen, J.H., Petersen, M.M. and Lund, B. (1993) "Effect of External Hip Protectors on Hip Fractures," *Lancet* 341: 11–13.

Michael, M. (1997) "Inoculating Gadgets Against Ridicule," *Science as Culture* 6: 167–193.

NIH (Nuffield Institute for Health and NHS Centre for Reviews and Dissemination) (1996) "Preventing Falls and Subsequent Injury in Older People," *Effective Health Care* 2: 1–16.

Piaget, J. (1932) *The Moral Judgement of the Child*, London: Kegan Paul, Trench, Trubner & Co.

Porter, T. (1986) *The Rise of Statistical Thinking 1820–1900*, Princeton: Princeton University Press.

Stone, D. (1991) "Preventing Accidents – A High Priority Target," *Medical Monitor* 24 May, 61–65.

2 Narratives of trust in constructing risk and danger

Interpretations of the Swissair 111 crash

Jörg Potthast

> As soon as he was able to recall the events in the sequence of their temporal appearance, he felt as if the sun shone on his stomach.
>
> (Musil 1952: 650)

In the fall of 2001, a number of dramatic airplane crashes shook the world. Four airplanes crashed on September 11, 2001 in New York City, Washington DC and Buckstown, Pennsylvania, as a result of terrorist acts. Soon after, another airline crash took place over the Black Sea (October 4), and a deadly collision between a large SAS plane and a small Cessna took place in Milan (October 8). Finally, New York City was struck again by an airplane disaster on November 12.

Three years earlier, on September 2, 1998, a Swissair airplane departing from New York with 229 people on board had crashed into the sea close to Halifax, Canada. There were no survivors. In its proportions, this crash was not comparable to the events of September 11, but it was immediately followed by an extensive debate in the press and on the Internet. The day after the accident, a *New York Times* Web Forum received some sixty letters concerning the crash. Within a month, more than 500 letters were sent to this Forum. This chapter offers a close reading of the comments expressed in these letters.

In addition to the enormous loss of civilian lives which all the crashes mentioned above have in common, the Swissair case was characterized by an extreme uncertainty as to what caused the accident and who is to blame for it. As a result, a large proportion of the Forum letters were concerned with presenting theories to account for the crash. Among contributors who disclosed their identities were flight engineers, pilots, pilot instructors, military pilots, sports pilots, anti-terrorist-specialists, Swissair company members, flight attendants, space engineers, physicists, frequent travelers, programmers, aviation amateurs and people living close to Halifax. The contributions differ in content and form. Some of them are well-composed letters, while others present fragmentary lists, probability calculations or quotations from available documents.[1]

Since both the flight data recorder and the cockpit voice recorder had stopped working 6 minutes before the plane dived into the sea, Internet exchanges quickly focused on the question of what happened during these crucial six minutes. In their attempts to answer this question, some contributors made highly selective use of the available data and were strongly accused by others of giving "irrational" explanations. In this chapter, following a detailed account of a series of explanations offered by participants, I will, however, avoid drawing a distinction between rational and irrational accounts. Instead I seek to understand the level of trust that contributors continue to express in the large technical system of air transportation – regardless of the character of their contributions. I will argue that this sample of crash theories can shed light on a number of questions that much psychometric research on "risk acceptance" and "trust in technology" fails to answer. These questions include: Under what conditions do people continue to have trust in technological systems? How do people restore their trust in technology? Does awareness of the inadequacy of expert knowledge affect lay trust in the system?

It is now commonly assumed that trust is a necessary precondition for the "lifting out of social relations from local contexts of interaction and their restructuring across indefinite spans of time–space" (Giddens 1990: 21). Yet the basis for public trust in such processes of disembedding remains quite unclear, in particular regarding the functioning of large technical systems. As a matter of course, influential sociologists have tried to integrate trust into theoretical frameworks of the workings of modern sociology (Luhmann 1968; Giddens 1990). Among others, they point out that personal trust has been transformed into system trust in the course of modernization. While the notion of trust has been well established in sociological discourse, it has tended to remain, however, a kind of upgraded residual category. This is why the *theoretical* discovery of trust did not contribute to opening the field for the *empirical* study of trust (Hartmann 2001). On the contrary, as long as trust continues to be considered in a functionalist perspective as a *mechanism* compensating for complexity (Luhmann 1968) or as a necessary complement to the diffusion of disembedding mechanisms (Giddens 1990), the question of how trust is produced, reproduced and maintained in actual practice does not even seem worth being asked.

Students of large technical systems and High-Reliability Organizations (HROs) have been attentive to the phenomenon of trust as an indispensable organizational resource in both high and low risk contexts (see Rochlin and Sanne this volume). With regard to processes of innovation, extension or recombination, it has been stressed that large technical systems need to be "culturally embedded" (Summerton 1994). Yet, what does that "cultural socket" of large technical systems (Gras 1997: 30) consist of?

Douglas and Wildavsky convincingly argued that changing patterns of risk acceptance can only be explained by a "cultural approach [which] can integrate moral judgements about how to live with empirical judgements about what the world is like" (Douglas and Wildavsky 1982: 10). On a smaller scale, Garfinkel's work on "repair work" following "discrediting events" is also relevant in explaining how trust is (re-)produced and maintained (Garfinkel 1967).

In this view, the Web Forum about the Swissair 111 crash is neither an indicator of trust nor distrust *per se* but reflects an extraordinary collective attempt to *restore trust*. Participants in the Forum do not act, as social scientists sometimes assume them to do, like "cultural" or "judgmental dopes" (ibid.: 66) who keep trusting in a silent and passive way; they are not patiently waiting for experts to tell them what really happened. The question for social scientists is how to analyze the trust work they do and how best to understand and represent their striking interpretive competences.

Trust has then to be actively constructed, and it is this active construction that usually takes place following disruptive experiences. By focusing on the *New York Times* Web Forum as an episode of trust reconstruction after the Swissair 111 crash, I suggest that the question of how trust is constructed can be reframed as a question of how interpretive activity unfolds and is expressed among participants. In Section I of this chapter, I will illustrate how initially, the limited available data were used by contributors to the Forum in a highly selective manner and how these data were received. The ways contributors selected and recombined available data are identified as approaches for broadly defending and reaffirming a modern world view. In Section II, more comprehensive explanations of the crash – based on more information – are analyzed. Although now much more explicit about their background assumptions regarding trust in technological systems than people are under normal circumstances, the Forum contributors do not converge towards a shared "neutral" description of what really happened. Instead, as I will argue in Section III, the interpretive strategies deployed by the participants can be understood as three distinct narratives of trust. Apparently, those who demonstrate trust in the system have different reasons to do so. What does this finding imply, that is, what does it mean that people share an urge to re-establish causality but fail to converge towards a single concept of causality?

I: Dealing with lack of information, making sense of the accident

[I]t was a terrorist . . . i think . . . that guy from Afghanastan.

(contribution #107)

(Author's note: according to the list of passengers both a high ranking diplomat from Saudi Arabia and a man from Afghanistan were on board the crashed plane.)

[I]slam in general has never gotten over the fact that us and britain gave palestine to jews after ww2. remember, islam holds palestine as its holy land. i don't think it was iran, i think it was islamic hardliners.

(#34)

Before you decide that this was a terrorist act, think! What terrorist in
his/her right mind would blow up a Swiss plane!? Switzerland is where
they keep their money and launder their funds. They are crazy, but not
stupid!

(#30)

What really did happen on September 2, 1998? Experts in charge of official investigations are usually confronted with incomplete data. In airplane
crashes, three sources of information have to be considered. A first set of data
is provided by the information recorded by the black boxes that are installed
in every airplane and designed to withstand crashes. Second, an "archaeo-
logical" analysis of the state and dispersion of on-site debris may lead to
conclusions concerning the crash. A third source of information is witnesses'
accounts. These three sources of information have to be cautiously combined
within a long process of interpretation. Typically, a final report is not forth-
coming within less than two or three years following the crash.

In the case of the Swissair 111 crash, many of the contributions to the
Web Forum were written before the airplane's black boxes were localized
and recovered. These contributions were, at least in some cases, based on the
following sources: a published transcription of communication between
the cockpit and flight control, the passenger list, information on the approx-
imate location of the crash, and some witnesses' reports published in the
press. However, the relative lack of information is compensated for by other
means (such as press reports), which leads some contributors to accuse others
of expressing mere speculation. Still other contributors explicitly argue in
favor of speculation, referring to examples of plane crashes that have been
explained thanks to the speculation of laymen:

Cases like these are strictly unique and therefore require intuitionist
approaches to be solved.

(#140)

Even if there was only one case that had been resolved due to intuitionist
speculation, there would be no reason to blame "instant experts".

(#67)

or "instant answers."

(#232)

Large technical artefacts of a high degree of complexity are beyond the
scope of reverse engineering.

(#137)

In turn, some contributors complain of "hurtful and ignorant comments
in complete disrespect for the dead" (#393) and blame those who spread
unsustainable accusations. "Don't blame Swissair, . . . don't blame the bogey

man" (#238). Even for pragmatic reasons, speculation and conspiracy theories should be excluded: as one contributor notes, in the case of TWA 800,[2]
the official investigation had to face so many private investigators that large
parts of their work were devoted to disproving a false "report" (#140). While
conspiracy theorists cannot be excluded from the Forum, they are labeled as
being irrational if not insane.

Save your modern world view by any means

Three days before the Swissair 111 accident, another airplane had crashed in
Quito, Ecuador. Like the Swissair accident, the Quito crash was reported in
numerous media in the German-speaking world. A synoptical analysis of a
number of daily newspapers in German demonstrates, however, interesting
aspects about the way the media reported this crash.

> At least 77 people were killed in a crash of a Cuban passenger plane in
> Quito, the Capital of Ecuador. The plane collided into a settlement just
> behind the airport. According to eyewitnesses, the aircraft, a Russian
> Tupolev-154 that is run by the Cuban airline Cubana, caught fire during
> the start. It failed to take off correctly and broke through the airport
> barriers at the end of the runway. The burning machine touched the roof
> of a garage, slithered over a highway and broke into parts on a soccer
> field . . . Quito airport is considered to be one of the most dangerous
> airports in the world. At the same place . . . several accidents have
> occurred before. Only two years ago, a Brazilian aircraft ran into the
> barriers, and in 1994, 65 people died when a plane failed to take off.
> (*Der Standard*, Wien 31.8.98)

A majority of the selected newspapers makes one story out of two catastrophic events.

> The rescue activities [for the Quito crash] were still going on when the
> radio began reporting on a second disaster. In a mountainous area south-
> east of Quito, a lorry with 50 people on board went off the road and fell
> into a 80 meter deep gulch . . . The chauffeur and the passengers were
> on their way home from a party.
> (*Stuttgarter Zeitung* 31.8.98)

As the Quito crash was seemingly overdetermined, precise reporting
appears to be unnecessary. Both the airplane and lorry accidents do not
require any further explanation but are simply reported together in order to
comment on each other. Whereas an enormous interpretive apparatus is
mobilized in search of the causes of the Swissair crash, the comments on the
Quito crash appear to have a different point of departure: given so many
causes, how is it possible that accidents do not occur more frequently?

In the accounts above, the progression of the Quito accident is told in many different ways. The stories consist of a series of parts that are deliberately combined and accumulated: "collided into a settlement," "touched the roof of a garage," "caught fire," "soccer field," "highway." Also, what could be the end of a story that begins with a "Russian aircraft of the Cuban airline Cubana, starting from an airport in Ecuador," if not a catastrophic one? If the accident was not due to a serious defect of the aircraft, the airport is implicitly seen as likely to be responsible for the disaster. (It can be noted, however, that a previous accident reported to have taken place at the Quito airport in 1994 turned out to be pure invention.)

> Quito airport is situated in the middle of a huge housing area and therefore has a reputation of being one of the most dangerous airports in the world . . . Originally, the airport was situated at the outskirts of Quito, but as a result of the tremendous growth of the capital, the runways are now surrounded by inhabited areas.
>
> (*Berliner Zeitung* 31.8.98)

Here the reader is expected to "get the picture." Quito airport is surrounded by a housing area and high mountains. Badly nourished children play soccer and five of them are hit by the burning aircraft (*Berliner Zeitung* 31.8.98). Meanwhile, a highly overloaded lorry falls into a gulch. For sure, the chauffeur and his passengers were not going to a party (*Berliner Morgenpost* 31.8.98) but were making their way home from a party (as sustained by *Stuttgarter Zeitung* 31.8.98). The people of South America live in illegally proliferating settlements around airports, they use airplanes that have been built several years ago in a country (Soviet Union) that does not exist anymore. They commit their lives to airlines that may not even be covered by insurance (*El Comercio Quito* 1.9.98), but they know how to party.

What is the relevance of media reports of the Quito accident for our understanding of the response to the Swissair 111 crash? Several authors of the Web Forum concerning the Swissair 111 accident agree that catastrophes that occur in the "Third World" cause much less sensation in these countries than do accidents in the Western world. Although such catastrophes take place far away from the Western world, there appears to be no shortage of explanations as to why these planes crash. Is it that remarkable that professional journalists from well-established newspapers draw on cultural stereotypes, sometimes obscuring details in order to create a story that is coherent with their world view?

When both airline catastrophes are compared, it turns out that a "Third World" crash can be much more easily "explained" than the Swissair case. Also, the comparison throws some light on a common pattern of explanation: many of the possible causes collected by the Forum authors are presented as "Third World causes." To fulfill this purpose, the definition of "Third World" is considerably extended:

I'm not an aviation expert but I do fly trans-Atlantic out of Kennedy at least twice a month and given the "Third World" face the airport puts on for passengers, I can't help but think that the maintenance and technical facilities are equally unimpressive.

(#151)

Here we go again. Another plane crashes that was going from Kennedy Airport or going to it . . . (i.e. Pan Am in '88 and TWA in '96).

(#44)

Kennedy Airport is not the only weak element within the global aviation system to be given a "Third World" connotation. Writers do not understand "why in today's 'modern world' it took about 3 hours to find the plane" (#8). The short but convincing answer was: "All corners of this world are not equally modern" (#28). Similarly to the area in which the Quito accident took place, the area where the Swissair plane came down is no "high-tech" region; but sparsely populated (#37). Obviously, it did not help that its inhabitants used to be simple fishermen and are said to have highly specialized forms of knowledge about the ocean:

They know the sea as few others do and they know that as devastatingly powerful as it is, it can also sometimes be fickle, sparing one while consuming the rest.

(#76)

Countless interpretations thus limit the range of possible causes to the existence of non-modern elements within the modern aviation system. Expressing and supporting this explanatory framework turns out to be more important than putting forward consistent arguments.

Distinguishing between expert and layperson accounts of the accident

The interpretations of the previous section confirm the culturalist view of Mary Douglas (1992) and others that there is no such thing as "real risk" as opposed to "perceived risk." Not only unidentified Web writers but also well-educated journalists who publish in established newspapers patently try to "save" their modern world view by any means. Danger might give rise to dark and metaphysical speculation, but in a modern world, there is only risk. Within modern society, risk – especially the risk of traveling by plane – is estimated to be extremely low (#41, #400).

The idea that risk is calculable and can be objectified by experts is, however, vividly defended on the Forum against those who challenge this idea.[3] Analysis of the Forum letters – as well as responses to these letters – shows that the contributors are keen to distinguish between expert and layperson accounts of the Swissair 111 accident. Contributors who are regarded as laypersons are challenged and labeled in various derogatory ways. Consider the following exchange, which starts with this contribution:

> Now we have the Dilemma: Should the plane dump its fuel first?
> The following is my analysis:
> 1. Dump the fuel first:
> Assume the plane would be safe at the end with 50%, i.e. 50% probability 229 victims (as it was)
> 2. Land first with overweight:
> Assume the 50% probability 229 people on board alive.
> i.e. 50% the plane crashed at the end, in this case, the plane may miss the runway and crash into the Halifax downtown,
> so the probability is 50% of 229 victims and 50% more than 229 (say 500).
> So, the estimated victims are as follows:
> 1. $0 \times 50\% + 229 \times 50\% = 114.5$
> 2. $0 \times 50\% + 229 \times 50\% \times 50\% + 500 \times 50\% \times 50\% = 182.5$
> Because if dump the fuel first, the estimated victims (on the plane and ground) is 114.5, while a directly landing without dumping the fuel may end up with an estimated victims to 182.5,
> Therefore it is justified to dump the fuel first.
>
> (#335)

The author of this contribution is immediately called a paranoid by other contributors (#336, #337); he is told to stop his stupid calculations and sharply recommended "to go real" (#340). Seemingly unimpressed, the author replies:

> Dilemma is Dilemma, all comments are welcome.
> I only use simple assumptions in my calculation, but I think I already pointed out the Dilemma.
> Should you have accurate data, please give me, I will calculate a more precisely probability for you.
> This time, we may need complexed equations, natural e, or even Pi.
>
> (#343)

Another issue that is discussed by some contributors is how the Forum should be protected against "fake experts":

> Having "Joe Pilot" with thousands of flight hours, and "former aviation safety Inspector Whomever", (usually, a paid network consultant) telling us what may have happened is for the seriously misinformed.
>
> (#140)

The appearance of fake experts does not, however, appear to weaken the contributors' trust in "real" experts. How then to qualify as a "real" expert under the particular conditions of a virtual Forum? In some contributions, there is a conscious effort to present crash theories in an "expert style." Contributions tend to be as data-based as possible, quoting eye witnesses (#199, #216) and several specified documents (#216, #402). Sometimes, the lists of technical details presented (#139, #225) are too long and contain too many abbreviations to be comprehensible (#373, #471).

To gain credibility and to be referred to as an expert is nevertheless hard work. Those whose opinion is solicited have to follow the discussion on the Forum continuously: it is interesting to observe how a small group of quasi-official experts regularly intervenes and gradually takes over the task of moderating the discussion. Together, "informed people" and "guesswork people (me included)" (#350) are assumed to have prepared the field for official experts who have to go further (#343). While not everyone congratulates the Forum for being an expert support movement, only a few voices explicitly cast doubt on the Forum's success as a parallel investigation:

[T]here is no way on this earth that a few disconnected minds on this Forum could contribute anything useful in the absence of concrete evidence.

(#371)

I am still sick of those second guessing the decisions made by the pilot, ATC, the designers in the COMPLETE ABSENCE OF ANY USEFUL INFORMATION.

(#400)

These contributors sound quite pessimistic, but because both have been written by a single author who is among the busiest and most constructive writers on the Forum (#114, 280, 347, 434, 445, 460), they have to be interpreted as an expression of individual skepticism. Other contributors enthusiastically report that a great source for understanding "the truth" has been discovered if only one knows how to use it:

Well, I have taken part in several discussions on NYT Forums. I have sifted through crap, hate mail, and bigotry, but in doing that I have encountered some very good wisdom, and found some people that share my care for all humanity. I have learned a great deal. I have learned stuff that would have taken me months of reading, because the good posters are either experts in some field or they have read and researched areas that I haven't gotten round to . . . You just have to read everything and balance it in your own mind. That's how you end up with the truth.

(#378)

Rather than affirming trust in experts, this author develops a guideline on how to restore trust in the absence of clearly distinguishable expert knowledge. Following his recommendation, I have identified two clusters of crash hypotheses which will be closely examined in the next part.

II: Making the most comprehensive use of available data

Contributions to the Web Forum concerning the Swissair 111 accident reflect two main causal hypotheses, namely an explosion hypothesis and an overweight hypothesis. Both of these hypotheses call for a precise chronological reconstruction of the catastrophic event. In the following, I will discuss each of the hypotheses in turn.

The explosion hypothesis is supported by the observation that the aircraft was broken into small parts of debris. There are various interpretations regarding the question of when exactly the explosion occurred. In order to fixate the exact time of the explosion and to answer the question of how it came about, the last conversation between the cockpit and air traffic control is quoted in its entirety:

—— BEGIN QUOTE ——
Controller: turn left to lose some altitude
SR 111: Roger, we are turning left.
SR 111: We must dump some fuel. We may do that in this area during descent.
ATC: Okay
SR 111: Okay, we are able for a left or right turn toward the south to dump.
ATC: Roger, turn left heading of two hundred degrees and advise me when you are ready to dump.
SR 111: We are declaring an emergency at time zero one two four we are starting vent now. We have to land immediately.
ATC: Swissair one eleven, you are cleared to commence your fuel dump on that track and advise me when the dump is completed.
ATC: Swissair one eleven check you are cleared to start fuel dump.
Note: There were no further communications from Swissair 111. Approximately six minutes later, the aircraft contacted the water.
—— END QUOTE ——

(#216)

Early interpretations of parts of this conversation had concluded that the plane crash was caused by lack of fuel (#16, #25). In response, others had replied that it would be impossible to dump too much fuel because of an automatic retain mechanism (#26, #161). Any hypothesis assuming that the crash was caused by lack of fuel would thus have to explain why this retain mechanism had stopped working (#26). Ultimately, the lack-of-fuel hypothesis is rejected when someone observes that it would have been impossible to dump too much fuel within 16 minutes (after the first emergency call) or even six minutes (after the permission to dump fuel) (#139, #209).

The explosion hypothesis is further developed when the shortcomings of the lack-of-fuel idea become apparent. The central argument is based on the assumption that there was a steady flow of conversation between the cockpit and air traffic control. Why did an ongoing conversation – running over a longer period of time without any considerable interruption – stop suddenly after permission had been given to dump fuel? The supporters of the explosion hypothesis assert that dumping the fuel must have caused an explosion, which in turn explains the sudden end of the conversation: the dumped fuel immediately caught fire and, catching up with the plane, led to an explosion of the fuel tanks. This hypothesis is consistent with the report of a witness who affirms that the engines of the plane continued to work for some time after the explosion. In the opinion of its supporters, those denying this hypothesis have to provide two independent, alternative causes for the explosion and for the power failure that led to the sudden end of the conversation between the cockpit and air traffic control.

The explosion hypothesis entails constructing and interpreting a very detailed account of what could have happened. Because of the need to integrate observations from different sources, its presentation is much longer (up to 1,000 words, #216–217) than the average contribution to the Forum (144 words). To reinforce his credibility, the principal author of the explosion hypothesis refers to his scientific authority, signing with his complete academic address as follows:

Jason A. Taylor Department of Physics NASA/GSFC Code 661 United States Naval Academy. Greenbelt, MD 20771 (USA) Annapolis, MD 21402–5026 (301) 277–1909

(#217)

The explosion hypothesis is not, however, the last hypothesis to be discussed on the Forum. It is succeeded by the overweight hypothesis, which has a more complex temporal structure and identifies a paradox which cannot be resolved by reducing the catastrophic event to a sequence of probable micro-events. The paradox is based on the fact that the maximum allowed take-off weight of a plane is usually higher than its maximum landing weight: in the period following its take-off, the Swissair plane is too heavy to successfully make an emergency landing. This overweight problem constitutes a decision-making dilemma for the pilots, which in turn is a source of controversy among the participants of the Forum.

This decision-making dilemma can be described as follows. At a distance of 30 miles from Halifax, the plane is still too heavy to make an emergency landing. For this reason the pilots decide to make a loop to dump the excess fuel over the Atlantic. Because the plane had some 30 tons of overweight

(#211), this decision is regarded by some Forum contributors as reasonable (#116, #120), also considering the fact that an emergency landing with full tanks is the worst alternative if there is a fire on board (#114). Opposed to this comment, it is claimed that there is no reason to waste time if there indeed is a fire on board (#179). Several contributors to the Forum are convinced that it is possible to land a plane at a weight of 10–20 percent (#475) "above the 'book' value" (#209). In any case, ten minutes of controlled flight after the first emergency announcement should have been enough time to bring the plane quite close to a safe landing weight (#403). According to these comments, the pilots, who thus wasted precious time, must have underestimated the seriousness of the situation (#264).

There is no consensus on the Web Forum concerning the overweight hypothesis. Its opponents insist that it is only possible to land an overweight plane if the plane is still under control. They further point out that the maximum landing weight also depends on the length of the runway and on weather conditions (#435). The supporters of the hypothesis reply that the pilot not only started too late to dump fuel but also lost too much time before reducing altitude. It is calculated that if he had reduced altitude immediately, the plane would have reached the ground within ten minutes (#403).

In order to save the plane and the lives of the more than 200 people on board, the pilot would have had to do two things that are essentially mutually exclusive, namely to dump fuel and to land. Should then the Swissair 111 accident be viewed as a result of pilot error or system failure? One contributor considers it a scandal that the maximum starting weight exceeds the maximum landing weight (#166). Given this indicator of system failure, it is perhaps surprising that no contributor to the Forum makes a clear political argument against system complexity as a basic cause of "normal accidents" (Perrow 1984).

The real shock: finding the black box empty

In light of the difficulty in reconstructing the chronological order of crash events, another question arises on the Forum, namely: how is it possible that we know so little about the disaster? This issue is rapidly politicized. Initially, some contributors to the Forum express considerable faith in the plane's black box. Trust in technology concentrates highly on the possibility of objective reconstruction of events:

> Once they have the black box, we will know every little detail of what happened. So let's wait.
>
> (#210)

Subsequently, many authors refuse to accept the fact that the black box failed to provide an unambiguous record of what really happened. More shocking than the crash itself seems to be the recognition of the impossibility of reconstructing it objectively. Who is to be held responsible for the absence of evidence, if not Clinton (#176, #178), the "canucks" (#396) or Islamic terrorists (#34, #263)?

An extensive discussion ensues. What is the problem with the black box, how was it possible for it to fail, and what could have been done to prevent such failure? On the Forum, suggestions are made to equip the black box with an extra battery (#257, #265). Other contributors maintain that such measures would not solve the problem:

> If no signals are given to the recorder, no music will come out.
>
> (#367)

> Even with extra batteries, a black box is useless when the plane is suffering an electric failure.
>
> (#276)

At this point, the discussion becomes increasingly aggressive. How is it possible that in the age of satellite communication systems, we still rely on black boxes (#156)? What about a simultaneous transfer of all relevant data to air traffic control centers instead (#157)? If it is possible to receive pictures from the moon, why do we still rely on Stone Age technology in this case (#164)? Generally a plane's black box is found within a few days after a crash, but "to have to wait for a Black Box to tell what happened is so unreal that it defies all common sense" (#164). One contributor wonders why we know why dinosaurs did not survive whereas we still ignore who has killed Kennedy (#38). Viewed as hopelessly technologically outdated, the black box is treated as a political artifact.[4]

Some contributors later express profound disappointment that the black box did not contain the data it was expected to provide:

> I believe we need answers, and my point earlier was that, hell, these black boxes are for the very purpose of recording those last minutes.
>
> (#287)

Obviously, some Web Forum contributors have regarded the black box to be the infallible core of a technological system. The idea of the black box is symbolically suggestive: it gives rise to the idea that there is something that resists contingencies even though it is embedded in a vulnerable technical environment. Yet at the same time, for many contributors the black box is not worthy of the trust that has been bestowed upon it. Instead of providing "objective" data that has not been contaminated by political or economic interests – or revealing something that would justify the contributors' faith in "data" (mentioned 77 times) or "facts" (mentioned 98 times) – the black box fails. Although it has been designed to make the facts speak for themselves, it remains silent. Opening the black box and finding it "empty" contributes to making the disaster dramatic: if only there had been a way to reconstruct events in their correct temporal sequence!

Trust as a condition of modernity has been dealt with as an issue of social delegation of knowledge: the less we know about the technicized environment we live in, the more we rely on experts whom we trust to know about the increasingly complex relationships between causes and consequences.[5] Hence an awareness of the inadequacy of expert knowledge should be expected to deeply affect lay trust in abstract systems. Trust defined as trust in the successful delegation of knowledge should end here and turn to distrust.

Yet it can be argued that despite the extensive requiems over the loss of objective chronology, the vast majority of Forum contributors continue their activities for restoring system trust. What remains intact is precisely the desire to (re-)sequentialize the events leading to the crash, and the common element among all the explanations is a shared effort to sequence the disaster. This pattern is a clear example of a narrative of trust: participants' share their competences to translate an absurd event of high interactive complexity into a sequence which makes sense and which implies a moral justification for how to act.

III: Three narratives of trust in collective interpretations of the accident

Interestingly, the suggestions that were discussed in Section II do not converge toward a single mode of sequencing complexity. Instead, the various contributions to the Forum concerning the Swissair 111 crash can be viewed as expressing three overriding narratives of trust.

Narrative #1: Improving technology, sequencing the complexity of events

As indicated above, the first narrative of trust reflects the stance that crises could be avoided or handled by improving the technology and sequencing the complexity of events. Applied to the Swissair 111 crash, this approach means trying to achieve an acceptable sequentialization by design: it should be possible to prevent accidents from happening by technologically insuring a more stable chain of events. Consequently, a number of contributors to the Forum make concrete suggestions for safety innovations such as emergency capsules, ejection seats and parachutes. If there is no way of preventing crashes from happening, technological devices should be in place to prevent people from dying. For example, assuming that there is usually sufficient time for passengers at least to put on parachutes and move towards emergency exits (#91), why not equip each passenger with a parachute (#126)? One contributor suggests an answer: in civil aviation, such measures have been abandoned for weight reasons (#138). In terms of weight allowances, one must choose between parachutes and hand luggage. Even if this problem could be resolved, a more serious one would still remain, namely how to avoid panic and chaos as passengers clamor toward the emergency exits (#98). One contributor suggests the following technologized solution:

> [T]o reduce panic after the bail-out offer, I thought of a device which would restrict co-motion in the plane. Electrically locking seatbelts through control by pilots may be one of the options. In an emergency, a pilot would command everyone to sit down, lock all seat belts, ask everybody to remain in their places, announce emergency, and offer a bail-out. Then, he could unlock row by row.
>
> (#126)

Another contributor suggests a "detaching chamber":

> But what if – and only what if – a detaching chamber, by means of rockets for instance, could take people safely to the ground without them having to struggle with individual parachutes, physical handicaps and (a big AND) other passengers trying to save their lives first?
>
> (#98)

Alternatively, modern space technology inspires some contributors to conceive of a passenger cabin that splits into several autonomous parts, leaves the aircraft with rocket drives and lowers itself softly to the ground (#71, #72). Others object to this proposition as unrealizable, pointing out that rocket driven modules or ejection seats cannot be used by everybody:

> Even with a modern ejection seat exiting the aircraft is no cake walk. And we are talking about young people in peak physical condition, not your grandmother.
>
> (#95)

Exit by ejection seats is viewed as unfit even for passengers in the best physical condition:

> Ejection from an aircraft is an extremely *violent* act. The human body is subjected to accelerations of up to 15 g's. For an instant, a passenger is 15 times his/her weight. The most modern rollercoaster doesn't exceed more than two g's.
>
> (#356)

In many cases, suggestions for technological innovations to improve passenger safety pose new questions. Why is it not possible to design aircraft that could make successful sea landings? As currently constructed, planes are too heavy and too fast (#286). While an alternative would be to make planes considerably smaller, lighter or sturdier, it is feared that such measures would totally "mess up" the organization of air traffic (#27). One contributor also notes the potential economic consequences of these types of solutions:

I won't hesitate to bet that it would impose such significant weight, performance and fuel consumption penalties that it would boost the cost of
the airframe and its operation well past the limits of affordable air travel.
(#104)

Many proposals thus lead to new awareness of the systemic nature of airline
transportation as a large technical system. Among others, contributors claim
that safety is not a matter of components but rather of system architecture:

MD builds a pretty good airframe – that's its schtick. GE, P&W, Rolls,
etc. all build great engines and power supplies. Sperry, Raytheon,
Motorola etc. make great avionics, etc. etc. What's left is "how do we
fit all these pieces together?" And the answer is: electric and hydraulic
connections. Get the picture.

(#252)

Many contributors agree that the main vulnerabilities of airplanes lie in
their electrical and hydraulic connections. This fact indicates a need to find
experts of "flange engineering" (#252) who would be capable of revolution-
izing traditional systems of compartmentalized development and production.
According to one contributor, the basic philosophy behind aircraft design
has to be reversed: instead of viewing an aircraft as an assemblage of parts,
it has to be approached as a system of interlinking connections. Hence, any
improvements in aircraft design and manufacture will have to start with
improvements in the training, existing knowledge and experience of manufacturing and maintenance staff (#410). Although today all central functions
of an aircraft are supported by various measures to insure redundancy within
the system, redundancy has its limits. Given the fact that there are already
260 kilometers of cables in modern aircraft, increasing the redundancy of
the system pushes maintenance to its limits (#400).

In conclusion, the various Forum proposals for increasing air safety through
improved technology entail such measures as designing optimal sequences
of events ("first parachutes, then eject"), improving airplane construction
("making aircraft lighter or seaworthy") and increasingly recognizing the
systemic nature of aircraft engineering. As such, these proposals constitute
a narrative of trust in the operation of the system. Perhaps not surprisingly,
however, what might be technically feasible often turns out to be socially
challenging if not impossible.[6]

Narrative #2: Strengthening social ties

The second narrative of trust that is expressed by contributors to the Forum
can be termed strengthening social ties. This narrative excludes technical
considerations such as sequencing as indicated above, while emphasizing
measures to enhance familial, community and social ties. Instead of

attempting to minimize the consequences of crashes as technical contingencies, the second narrative of trust has a retrospective aim of offering comfort and support to crash victims.

Nearly every contributor to the Forum who offers condolences supposes that the crash has, above all, destroyed families. On board the plane, there were "devoted mothers, fathers, tender babies, precious sisters and brothers from a score of countries" (#87). Only strong familial and social ties can provide support to those who lost parts of their families and whose pain cannot be financially compensated (#253, #256):

> We as a community must support the families.
>
> (#1)

The Swissair 111 crash thus recreates and underscores the importance of basic social relations. When technology fails and insurance cannot compensate, families and friends are needed. In contrast to the chronology of machines, the chronology of families is not linear; instead it is constituted by repetition and memorization of a number of rituals. Whereas a high number of contributors refer to "families" (the word "family" is mentioned 100 times on the Forum), there is only one contributor who expresses faith in the much more abstract concept of mankind:

> Mankind has reached the point of going to Mars and the Moon, Mankind deploys utterly disproportionate investments in searching ways to destroy and conquer one another. Mankind is pretty successful at this . . . But Mankind also has this beautiful power of putting his skills to the service of Mankind towards a "better" world and I have faith in Mankind.
>
> (#71)

A year after the disaster, with the official investigation still far from being completed, a ceremony is organized close to the crash site to commemorate the victims. A German newspaper reports:

> Canada is mourning with the surviving dependants of the disaster . . .
> A three day celebration of commemoration at the crash site ended Thursday night with a ceremony in Halifax. Under a starry sky, a ship's bell reminded of the moment when, one year ago, a MD-11 crashed into the Atlantic.
>
> (*Tagesspiegel* 4.9.99)

The celebration is a success. The disastrous events are inscribed in various ways within different religions and cultures, and a new "family" is created:

> As a multicultural immigration country, Canada had organized a closing ceremony in the town of Halifax where weeping people of all religions

came together with songs and prayers. An Imam was singing, a Salvation Army band was playing, and a Jewish musician was enthusiastically received by the community. Two blond twin girls from Sweden who had lost their father recited a poem they had composed themselves: . . . It seems as if an international community of solidarity has been created, promoted by the prime minister Jean Chrétien: "Consider the Canadians to be your family and Canada to be your home", he said to the mourners. "Merci, merci – I love you", replied one of the spokespersons of the bereaved families.

(Tagesspiegel 4.9.99)

The experience of discontinuity and loss thus entails a reaffirmation of social ties, and the commemoration of the accident is celebrated as the foundation of a new community. In terms of chronological time, it is impossible to undo the accident. However, it can be argued that it became part of collective memory, the crash had been successfully brought to a second type of narrative of trust, where rather than improved *technological sequencing* the emphasis is placed on *ritual sequencing*. As an effect, social, community and familial ties are strengthened.

Narrative #3: Improving operating conditions

Very simplistically, design of airliners is a juggling act where the balls are weight, revenue generating payload, and safety. How safe can we make an airplane without increasing the weight and hence decreasing the revenue generating payload?

(#356)

As the above contributor to the Forum implies, the metaphor of the juggling act can be extended to describe the temporally precarious condition of operating large technical systems. The third narrative of trust among contributors is thus concerned with improving operating conditions for such complex technologies as modern airplanes. One of the discussions on the Forum concerns the role of checklists in allowing, at least in principle, the operation of complex technical systems in a sequential mode (#154). As checklists have to be kept concise in order to be comprehensible in the case of an emergency (#410), this presupposes an abridged order which remains necessarily incomplete: how can the complexity of an aircraft possibly be so reduced and still remain intelligible? Several participants confirm that the pilots have followed the "Smoke/fire of unknown origin checklist" (#373, #388). Another contribution agrees:

If those guys were dumping fuel in an effort to lighten the aircraft before landing, they were quite likely following checklist emergency operating procedure for the MD-11 aircraft.

(#438)

The pilots are said to have paid too much attention to the checklist, a contributor suggests:

> In those last 16 minutes it was the decision to follow procedure by the book that cost us all 229 wonderful people. The procedures don't take into account the many variables that come into play in real world situations.
>
> (#121)

> Instead, I'm suggesting that that checklist emergency operating procedure was flawed, and this erroneous procedure may be traced back to whoever wrote and approved the emergency landing checklist.
>
> (#438)

Given these problems, reinstituting a third pilot is called for: three crew members and a clear separation of tasks are viewed as appropriate responses to disasters such as this one (#16). In a situation of emergency, there is no time to be lost by processes of social coordination, and any mode of organization among cockpit staff that might interfere with the ability to coordinate actions quickly has to be avoided:

> Here is how we solve this problem: never get into an airplane where BOTH of the pilots are instructors! Someone needs to be in control.
>
> (#277)

Another contributor suggests that while pilot instructors could be over-reflective, military pilots are used to operating within unquestioned hierarchies. Therefore the following measures should be taken:

1 Only allow military retired pilot to fly passengers plane, they are better trained and have a sense of flight.
2 A pilot would be forced to retire after 4 years of service, just like the US Presidency. This will prevent the pilots to take grant with anything by his basic instinct, instead of thinking by his brain.

> (#272)

This comment suggests trust in pilots possessing a sense of flight, but a lack of trust in pilots who depend on their basic instincts. There is an obvious tension between these two perspectives which makes it difficult to act on both of them simultaneously. Also, if emergency situations call for actions that are *both* based on routines and guided by reflection, how can a combination of these types of actions be achieved?[7]

Such questions are not possible to address through the various nodes suggested by the previous narratives of "technological fixes through sequencing" or strengthening social ties. The third narrative of trust is different from these other narratives. Instead, its point of departure is the actual operating conditions of those people (particularly pilots) who operate the large technical systems of air transportation. The contributors that express this type of narrative thus conclude that in the absence of any external guarantee for the temporal stability of these systems, the operating conditions of complex large technical systems must be improved.

Conclusion

Reports on smaller airplane crashes are usually placed in the "miscellaneous" section of most newspapers. In contrast, the press coverage in the aftermath of major aircraft disasters is typically quite extensive. In addition to front page coverage, background reports including technical details and analyses of comparable disasters are delivered on many pages and sections of today's newspapers. Business pages evaluate the economic consequences for airlines, insurance companies and sometimes even national economies. Magazine articles may take airplane crashes as an opportunity to reflect upon the modern condition and its myths and risks. Finally, in their media sections, newspapers sometimes reflect on the way disasters have been covered in other parts of the press. After the crashes of September 11, 2001, entire issues of newspapers were devoted to the exceptional, catastrophic events.

Major aircraft crashes are thus significant societal events that many people seek to interpret and understand, as indicated by the many contributions to the Web Forum discussed in this chapter. The approach here has been to select and analyze a limited corpus of material, guided by an interpretive principle of not distinguishing between "normal"/ "abnormal" or "rational"/ "irrational" contributions. Instead I have suspended judgments on the normality of the contributors. Also, any preliminary classification of the contributions would have misrepresented the level of uncertainty the interpreters of the crash experienced. In analyzing the material, the concept of *narratives of trust* has been developed as a strategy to reduce the diversity of the Forum's contributions, while also conveying differences in contributors' approaches to interpreting and explaining the crash.

The Web Forum concerning the Swissair 111 airplane crash is an interesting object of study because it shows how trust is restored following a major catastrophe. Viewed as a spontaneous and self-organized exercise in the public understanding of technology, the Forum reveals on what grounds narratives of trust are generated. The question of trust can thus be reframed as a matter of actors' competences to restore trust. In this perspective, the concept of trust promises to lend itself more usefully to empirical investigation.[8] Asking what it is that supports the extraordinary continuity of the world, Giddens comes closer to this type of paradigm shift in the study of trust than indicated earlier; he can indeed be read as being more concerned about "trust society" than "risk society" (Wagner 1994: 145).[9]

Searching for an explanation of people's trust, I have adopted a strategy that could be called a "hermeneutics of risk." In doing so, I subscribe to the "cultural turn" within social studies of risk according to which there is no such thing as "real risk" as opposed to "perceived risk" (Douglas and Wildavsky 1982). As a result, it is clear that people have notably different reasons to trust in technology. The three narratives of trust identified in Forum contributors' (re)constructions of the Swissair 111 crash reflect participants' views on how trust in the technological system should be restored and handled. These are (1) an approach of restoring or perfecting the system through technical improvements and enhanced sequencing of complexity ("technological fix"); (2) an approach of retrospectively strengthening social, community and familial ties; and (3) an approach of improving the operating conditions of these who work within large technical systems, particularly pilots.

If risks can be viewed as stemming from a plurality of heterogeneous cultural frameworks, the concept of risk cannot be reintroduced as an objective category of analysis. To escape such a version of "objectivism within relativism," I argue in favor of a *limited pluralism* by identifying a plurality of narratives of trust that are limited in number but share some common traits and concerns. The narratives show how the Forum, while reflecting a collective struggle to deal with the discontinuity and loss of the disaster, also ultimately restored trust in the system.

Notes

1   Once I had discovered the Forum, I was immediately caught up in its dynamic. I followed the discussion in "real time" but never joined in. Since I was about to start a PhD thesis on air traffic infrastructures, my situation was similar to that of the participants of the conference on "social responses to large technical systems" in Berkeley 1989: "Our welcoming dinner began about twenty minutes after a major earthquake struck nearby . . . One session the next day was spent comparing videotapes of the media response immediately after the quake with reporting a day later" (La Porte 1991: vii). Far from presenting any kind of "final crash report", this chapter is about early interpretative struggles to understand the Swissair 111 accident. The corpus of contributions is available at http://www.wzberlin.de/met/joerg_pub_en.htm.

2   TWA 800 crashed on July 17, 1996. Whereas official investigators have not yet found an "originating cause" either for TWA 800 or for SA 111, Elaine Scarry, a professor of English literature, has brilliantly argued for electromagnetic interference as a cause for both accidents (Scarry 1998, 2000; Eakin 2000).

3   This is exactly what Mary Douglas experienced when she presented a "theory of danger" which treated modern and pre-modern reactions to misfortune within a single framework. She reports that the professionals of risk assessment refused to acknowledge that the job they were doing had any political implication (Douglas 1992: 1–21). Many students of risk have followed Douglas in her ambition of uncovering political uses of "risk as a forensic resource" (Douglas 1990; Oliver-Smith 1996), though not all of them follow her culturalist model but subscribe to the interest model of risk acceptance in order to "denaturalize disasters" (Klinenberg 1999).

4   Is the black box really that outdated? Black boxes are about the size of a shoe box and are painted orange. They are designed to be fireproof, resist up to 1,100°C as

well as high impact (max. 3,400 Gs/6.5 ms) and high pressure (max. 6,000 meter deep water). When a black box touches water, a transmitter is automatically activated that enables localizing the box up to 4,200 meters depth in water. In the case of an accident, the black boxes are immediately transferred to the respective aviation authority when found. The microphone of the cockpit voice recorder (CVR) is installed in the middle of the cockpit console. Via analogous or digital recording a large array of noises is registered. "Sounds of interest to an investigator could be engine noise, stall warnings, landing gear extension and retraction, and other clicks and pops. From these sounds, parameters such as engine rpm, system failures, speed and the time at which certain events occur can often be determined. Communications with Air Traffic Control, automated radio weather briefings, and conversation between the pilots and ground or cabin crew are also recorded" (http://www.ntsb.gov/aviation/CVR_FDR.htm). Besides acoustic recording, the flight data recorder (FDR) is designed to save between 28 and 300 items of information concerning time, altitude, speed, course and location of the aircraft. In the case of an accident, these data are put together to produce computer-aided video animations of the flight.

5   As Green notes, "Weber could characterise the rationality of modernity as one in which all things were in principle calculable: as individuals we may not know how a particular technology . . . works, but we assume that someone (an expert) does know" (Green 1997: 132).

6   With regard to the case of air transport, students of large technical systems have demonstrated how the running of large technical systems implies complicated processes of social co-ordination (Gras *et al.* 1994).

7   Studies on accidents within large technical systems have been very attentive to this question. See Weick (1990a) on the Tenerife air disaster or the numerous accounts for the *Challenger* case. Re-reading a number of accident accounts written by social scientists, Pinch points out what makes this question so puzzling and the answers so inconsistent: there is no way to assign modes of organization to different areas of knowledge (Pinch 1991: 154).

8   Also, there is no more need to draw a sharp line between the general public passively expressing "attitudes" of trust or distrust toward risky technologies and the operators whose handling of technology involves active forms of trust. Both are ways of sense making (Weick 1995). According to Weick, technical systems consist of two intercalating layers, "technology on the floor" defined as materialized technical processes and "technology in the head" which refers to the representations of its users (Weick 1990b). In order to extend Giddens' concept of trust in expert systems, Schulz-Schaeffer redevelops Weick's argument to state that interactive complexity is not a property of technical artifacts but a result of decoupling processes between both layers (Schulz-Schaeffer 2000: 349–363).

9   Giddens writes, for example: "For basic trust in the continuity of the world must be anchored in the simple conviction that it will continue, and this is something of which we cannot be entirely sure . . . Business-as-usual, as I have pointed out, is a prime element in the stabilising of trust and ontological security, and this no doubt applies in respect of high-consequence risks just as it does in other areas of trust relations" (Giddens 1990: 146–147). Giddens frequently refers to air traffic to illustrate his argument on trust as a necessary complement of disembedding mechanisms (Giddens 1990: 29, 35, 112). With regard to airplane accidents, we regularly experience that even experts do not "know" why a technology has failed. Thus if "basic trust" persists, this points to a (narrative) mode of knowing in its own right which is not to be regarded as a deficient version of logo-scientific knowing (Czarniawska 1997: 17–21).

References

Czarniawska, B. (1997) *Narrating the Organization. Dramas of Institutional Identity*, Chicago: UP.

Douglas, M. (1990) "Risk as a Forensic Resource," *Daedalus* 119(4): 1–16.

—— (1992) *Risk and Blame. Essays in Cultural Theory*, London: Routledge.

Douglas, M. and Wildavsky, A. (1982) *Risk and Culture*, Berkeley: California UP.

Eakin, E. (2000) "Professor Scarry has a Theory," *New York Times* (19 November).

Garfinkel, H. (1967) "Studies on the Routine Grounds of Everyday Activities," in H. Garfinkel (ed.) *Studies in Ethnomethology*, Englewood Cliffs: Prentice-Hall.

Giddens, A. (1990) *The Consequences of Modernity*, Cambridge: Polity.

Gras, A. (1997) *Les macro-systèmes techniques*, Paris: PUF.

Gras, A., Moricot, C., Poirot-Delpech, S. and Scardigli, V. (1994) *Face à l'automate le pilot, le contrôleur et l'ingénieur*, Paris: Sorbonne.

Green, J. (1997) *Risk and Misfortune. The Social Construction of Accidents*, London: UCL Press.

Hartmann, M. (2001) "Einleitung," in M. Hartmann and C. Offe (eds) *Vertrauen. Die Grundlage des sozialen Zusammenhalts*, Frankfurt/Main: Campus.

Klinenberg, E. (1999) "Denaturalizing Disaster: A Social Autopsy of the 1995 Chicago Heat Wave," *Theory and Society* 28: 239–295.

La Porte, T. (ed.) (1991) *Social Responses to Large Technical Systems*, Dordrecht: Kluwer.

Luhmann, N. (1968) *Vertrauen. Ein Mechanismus der Reduktion sozialer Komplexität*, Stuttgart: Enke.

Musil, R. (1952) *Der Mann ohne Eigenschaften*, Reinbek: Rowohlt.

Oliver-Smith, A. (1996) "Anthropological Research on Hazards and Disasters," *Annual Review of Anthropology* 25: 303–328.

Perrow, C. (1984) *Normal Accidents. Living with High-risk Technologies*, New York: Basic Books.

Pinch, T. (1991) "How Do We Treat Uncertainty in Systems Failure? The Case of the Space Shuttle Challenger," in T. La Porte (ed.) *Social Responses to Large Technical Systems*, Dordrecht: Kluwer.

Scarry, E. (1998) "The Fall of TWA 800: The Possibility of Electromagnetic Interference," *New York Review of Books* (9 April).

—— (2000) "Swissair 111, TWA 800, and Electromagnetic Interference," *New York Review of Books* (21 September).

Schulz-Schaeffer, I. (2000) *Sozialtheorie der Technik*, Frankfurt/Main: Campus.

Summerton, J. (1994) "The Systems Approach to Technological Change," in J. Summerton (ed.) *Changing Large Technical Systems*, Oxford: Westview.

Wagner, G. (1994) "Vertrauen in Technik," *Zeitschrift für Soziologie* 33(2): 145–157.

Weick, K. (1990a) "The Vulnerable System: An Analysis of the Tenerife Air Disaster," *Journal of Management* 16: 571–593.

—— (1990b) "Technology as Equivoque: Sensemaking in New Technologies," in P. Goodman, L.S. Sproull *et al.* (eds) *Technology and Organizations*, San Francisco: Jossey-Bass.

—— (1995) *Sensemaking in Organizations*, London: Sage.

3 Resituating your data
Understanding the human contribution to accidents

Sidney W.A. Dekker

Nights are dark over Colombia, and December 20, 1995 was no exception. Clouds had washed up against the hillsides of the valley west of the Andes mountain range, while a passenger jet made its way down from Miami, Florida, to Cali, a provincial Colombian town more than 200 kilometers southwest of the capital Bogotá. The jet had about 15 minutes to go when the controller at Cali airport asked the pilots if they could land on a runway that lay straight ahead of them, instead of looping around the airport and landing in the opposite direction as the pilots had originally planned. The pilots accepted and prepared the flight computers (that really fly the jet) to take it to a navigational beacon straight ahead, close to the runway they were now headed for. Reading the code for the navigational beacon "R" off their flying charts, they typed "R" into their computer systems. Instead of flying straight ahead, however, the computers instructed the aircraft to fly to a point near Bogotá. This instruction made the aircraft swing to the left and into the hills on the east side of the valley. The jet impacted a hill close to the ridgeline, broke apart and continued to the other side of the ridge, killing 165 people. Only months after the accident, when critical computer components had been found, cleaned and reassembled, did investigators discover that there had been a mismatch between the paper chart in the cockpit and the database in the computer. "R" was really the code for a beacon near Bogotá, not the beacon near Cali.

There are always multiple stories about any single accident: a complex and terrible event like this one is able to support multiple interpretations of reality. Table 3.1 reveals two such different accounts of the accident. The first account, constructed by the Colombian authority, chooses events related to pilot performance (which is related to the fact that the Colombian authority employs the controller involved in the accident). The other account, constructed by the airline, focuses on factors that made life difficult for the pilots during their flight, such as the over-reliance on automation as sold by the industry, the language difficulties incurred in interaction with a Spanish speaking controller, and so forth.

Table 3.1 Different statements of cause about the same accident. Compiled from Aeronautica Civil (1996) and Flight Safety Foundation (1997)

The Colombian authorities say:	*The airline says:*
The Cali approach controller did not contribute to the cause of the accident	Approach control clearances were not in accordance with ICAO standards. The approach controller's inadequate English and inattention during a critical phase of the approach were causal
... and among the causes are:	*... and among the causes are:*
Inadequate use of automation	Inadequate computer database
The lack of pilot situation awareness regarding terrain and navigation aids	Lack of radar coverage over Cali
Failure to revert to basic navigation when computer navigation created confusion and workload	Manufacturer's/vendor's over confidence in flight computer technology and resultant influence passed onto pilots regarding computer capabilities
The flight crew's failure to adequately plan and execute the approach to runway 19	Failure of those responsible to ensure that computer database matched industry standards
Ongoing efforts to expedite the approach to avoid delays	The crew's task overload caused by unexpected change in assigned runway

These accounts reflect different investigations and different conclusions. One reason that multiple investigations of complex accidents point to so many different causes is that the systems that are vulnerable to failure are so well protected against such failure. A lot needs to go wrong for a system to be pushed over the edge to breakdown (e.g. Reason 1997). A great many factors – all necessary and only jointly sufficient – contribute to producing the accident. This means that when investigators trace a failure chain, the causal web quickly spreads out like cracks in a window. The pattern of contributions to any complex system failure is dense, and "primary" or "root" causes that are found are arbitrary choices or constructs by the one doing the looking. The example above suggests that the selection or construction of cause is determined in large part by organizational or sociological factors that lie beyond the actual events.

One interpretation of the Cali accident, based largely on the official investigation conducted by the Colombian authority, emerged two years after the accident. It concluded that the flight crew, in a hurry after a last-minute runway change, had lost their situation awareness and crashed into mountainous terrain (Flight Safety Foundation 1997). Such accounts are not just conclusions. By necessity they are assertions about a reality that surrounded and was experienced by the pilots that night. But what are the sources and processes through which such a reality gets reconstructed by subsequent investigators? Many accounts consistently reflect the hindsight bias – that is, the tendency to overestimate what the people involved in the accident could have known in advance, to see a complex, twisted history as more or less linear and inevitable, and to confuse one's own reality with the one that was unfolding around the involved people at the time. Indeed, evidence of the hindsight bias in the interpretation above is piqued by the various psychological claims.

As this chapter will discuss, "Losing situation awareness" has little operational meaning other than indicating that we now know more about the pilots' situation (for example, where they were versus where they thought they were) than they themselves apparently did. Similarly, "hurry" is not a product of the data itself, it is only a construct, an investigatory leitmotiv to couple a late take-off in Miami with a wreckage on a hillside in Colombia. A "last-minute" runway change, similarly, is last minute only in hindsight when the outcome is known (and the hindsight bias also makes us believe that bad outcomes are necessarily preceded by bad decisions). When a jet has 15 minutes of remaining flying time, a cruising altitude of about 4,000 meters, and about 100 kilometers to its destination, a runway change cannot be construed as being "last minute" by any stretch of the retrospective imagination. But in interpretations of reality after an accident, historical accuracy is not the criterion that matters, as Weick (1995) reminds us, instead, plausibility is. Plausible accounts are convincing not because they are true (to paraphrase Nietzsche) but because they are convincing – even if they make lousy history (Weick 1995). Both lists in Table 3.1 choose and pillage pieces from context; they desituate performance fragments by excising them away from the circumstances that produced them and then rank these fragments in a way that generates an account that is merely plausible to the audience for which it is intended.

This chapter explores reasons and mechanisms behind the desituating of data in aircraft accident analysis, including retrospection, the social construction of cause, and investigators' susceptibility to lay interpretations. In particular, two patterns in desituating data are identified, namely micro-matching and cherry-picking. By micro-matching is meant linking behavioral fragments with after-the-fact-worlds. In cherry-picking, the investigator gathers fragments from a wide variety of places and times in the accident record – not because these fragments necessarily belong together but because they all support an argument that the investigator wants to make. Since the reasons and mechanisms behind such desituating of data stem in part from conventional and technological restrictions on the gathering and analysis of data, the chapter starts with a brief discussion of the technical context that surrounds this type of analysis. It will then be shown that current approaches to aircraft accident analysis also have to do with reactions to failure, by which investigations often become retrospective, proximal, counterfactual and judgmental. The chapter also analyzes another reason for the desituating of data, namely the lack of agreed-upon methods for matching context-specific details of a complex behavioral sequence with articulated, theory-driven models of human performance. Instead, investigators typically rely on inarticulate lay interpretations that often make broad, unverifiable psychological assertions about the observed particulars of an accident. The chapter concludes by presenting steps that investigators could take to resituate human performance data in the reality as it was evolving around their subjects at the time.

Technical context of aircraft accident analysis

Developments in digital and other recording technologies have enlarged our ability to capture human performance data. In commercial aviation, the electronic footprint that any incident or accident leaves can be huge. Recent initiatives such as the formation of a Future Flight Data Committee in the United States, as well as regulatory proposals to lengthen the recording times on cockpit voice recorders testify to the value that is ascribed to accessible chronicles of human behavior (e.g. Baker 1999). But capturing human factors data addresses only one aspect of the problem. Our ability to make sense of such data and to reconstruct how humans contributed to an unfolding sequence of events may not have kept pace with the increased technical ability to register traces of such behavior. In other words, the growing dominance of human factors in incidents and accidents (Boeing 1994) is not matched by our ability to analyze or understand the human contribution for what it is worth (McIntyre 1994).

Data used in human factors accident analysis often come from a recording of human voices and perhaps other sounds (such as ruffling charts, turning knobs). These data can be coupled, to a greater or lesser extent, to how the process unfolded or how the system behaved over time. A voice trace, however, represents, only a partial data record. Human behavior in rich, unfolding settings is much more than the voice trace it leaves behind. The voice trace always points beyond itself to a world that was unfolding around the practitioners at the time – to tasks, goals, perceptions, intentions, thoughts and actions that have since evaporated. But aircraft accident investigations are formally restricted in how they can couple the recording of the cockpit voices to the world that was unfolding around the practitioners (in the form of instrument indications, automation mode settings and the like). The International Civil Aviation Organization's Annex 13, the governing document for aircraft accident investigations, prescribes how only those data that can be factually established may be analyzed in the search for cause. This provision often leaves the recording of the cockpit voices as the only factual, decontextualized and impoverished footprint of human performance. Making connections between on the one hand, the voice trace and on the other hand, the circumstances and people in which it was grounded quickly falls outside the pale of official analysis and into the realm of what many would call inference or speculation.

Apart from the formal provisions of Annex 13, the problem is complicated by the fact that current flight data recorders often do not capture many automation-related traces, that is precisely those data that are of immediate importance to understanding the problem-solving environment in which most pilots today carry out their work. For example, flight data recorders in many highly automated passenger jets do not record which ground-based navigation beacons were selected by the pilots, what automation mode control panel selections on such things as airspeed, heading, altitude and vertical speed were made, or what was shown on the both pilots' moving map displays. As pilot work has shifted to the management and supervision of a suite of automated resources (Billings 1997; Dekker and Hollnagel 1999), and problems leading to accidents increasingly start in human machine interactions, this lack of technical data represents a large gap in our ability to access the reasons for particular human assessments and actions in cockpits.

Situated cognition: but decontextualized investigations

The inability to make clear connections between behavior and local world hampers the analysis of the human contributions to a cognitively noisy, evolving sequence of events. Basic findings from cognitive science and related fields continually stress how human performance is fundamentally embedded in, and systematically connected to, the situation in which it takes place (Neisser 1976; Gibson 1979; Winograd and Flores 1987; Varela *et al.* 1995; Clark 1997). Human actions and assessments can be described meaningfully only in reference to the world in which they are made (Suchman 1987; Winograd 1987); they cannot be understood without intimately linking them to details of the context that produced and accompanied them (Orasanu and Connolly 1993; Woods *et al.* 1994; Hutchins 1995; Klein 1998). Despite this balance of scientific opinion (e.g. Willems and Raush 1969), critical voice traces from aircraft mishaps are still basically left to be examined outside the context that produced them and in which they once carried meaning. As a result, investigators easily confuse their own reality with the one that surrounded the practitioners in question. There appear to be two broad patterns in this confusion, that is, two ways in which human factors data are taken out of context and given meaning in relation to after-the-fact-worlds. The section that follows discusses these two patterns of desituating data.

Desituating data by micro-matching: holding individual performance fragments against the background of a world we now know to be true

Faced with complex and temporally extended behavioral situations, a dominant tactic used in aircraft accident analysis is to restrict the situation under consideration by concentrating on only one isolated fragment at a time (Woods 1993). Thus an incident or accident sequence is typically parsed up into individual, controversial decision points that can then – isolated from each other – be scrutinized for accuracy or reasonableness. Rather than understanding controversial fragments in relation to the circumstances that brought them forth, as well as in relation to the stream of preceding and succeeding behaviors which surrounded them, the performance fragment is held up against a world investigators now know to be true. The problem is that the after-the-fact-world may have very little relevance to the actual world that produced the behavior under investigation. The behavior is contrasted against the investigator's reality, not the reality surrounding the behavior in question. There are various ways in which such after-the-fact-worlds can be brought into being, of which three will be mentioned here.

First, individual fragments of behavior are frequently contrasted with written guidance that in hindsight can be found to have been applicable. Compared with such written guidance, actual performance is often found wanting: it does not live up to procedures or regulations. For example, "One of the pilots . . . executed (a computer entry) without having verified that it was the correct selection and without having first obtained approval of the other pilot, contrary to AA's procedures." Aircraft accident analyses invest considerably in organizational archeology so that they can construct the regulatory or procedural framework within which the operations took place. Inconsistencies between existing procedures or regulations and actual behavior are easy to expose when organizational records are excavated after-the-fact and rules uncovered that would have fit this or that particular situation. This type of analysis is not, however, very informative. There is virtually always a mismatch between actual behavior and written guidance that can be located in hindsight (Suchman 1987; Woods *et al.* 1994). Pointing that there is a mismatch sheds little light on the *why* of the behavior in question. It can also be pointed out that mismatches between procedures and practice are not unique to mishaps (Degani and Wiener 1991).

A second way to construct the world against which to analyze individual performance is to point to elements in the situation that were not picked up by the practitioners (Endsley 1999), but that, in hindsight, proved critical. Take the turn towards the mountains on the left that was made just before an accident near Cali in 1995 (Aeronautica Civil 1996). What should the crew have seen in order to notice the turn? According to the manufacturer of their aircraft, they had plenty of indications:

> Indications that the airplane was in a left turn would have included the following: the EHSI (Electronic Horizontal Situation Indicator) Map Display (if selected) with a curved path leading away from the intended direction of flight; the EHSI VOR display, with the CDI (Course Deviation Indicator) displaced to the right, indicating the airplane was left of the direct Cali VOR course, the EaDI indicating approximately 16 degrees of bank, and all heading indicators moving to the right. Additionally the crew may have tuned Rozo in the ADF and may have had bearing pointer information to Rozo NDB on the RMDI.
>
> (Boeing 1996: 13)

This example reflects a standard response after mishaps, namely to point to the data that would have revealed the true nature of the situation (Woods 1993). Knowledge of the "critical" data comes only with the omniscience of hindsight, but if data can be shown to have been physically available, it is assumed that it *should have* been picked up by the practitioners in the situation. The problem is that pointing out that it should have does not explain why it was *not*, nor why it was *interpreted differently* back then (Weick 1995). There is a dissociation between data availability and data observability (Woods 1993) – that is, between what can be shown to have been physically available and what would have been observable given the multiple interwoven tasks, goals, attentional focus, interests and – as Vaughan (1996) shows – the culture of the practitioner.

A third way of constructing after-the-fact-worlds concerns standards that are not obvious or specifically documented. These standards are often invoked when a controversial fragment (such as a decision to accept a runway change (Aeronautica Civil 1996)) reflects no clear preordained guidance but relies instead on local, situated judgment. For these cases there are always "standards of good practice" which are based on convention and putatively practiced across an entire industry. One such standard in aviation is "good airmanship," which, if nothing else can, will explain the variance in behavior that had not yet been accounted for.

These three ways of reconstructing causes of accidents illustrate how investigators micro-match controversial fragments of behavior with standards that seem applicable from their after-the-fact position. The problem is that these after-the-fact-worlds may have very little relevance to the circumstances of the accident sequence, and that the investigator has substituted his own world for the one that surrounded the practitioners at the time.

Desituating data by cherry-picking: grouping similar performance fragments under a label identified in hindsight

The second pattern in which meaning is imposed on available data from the outside and from hindsight consists of grouping individual fragments of human performance that prima facie represent some common condition. Consider the following example, where diverse fragments of performance that are not temporally co-located but spread out over half-an-hour are lumped together to build a case for haste as the explanation for the (in hindsight) bad decisions taken by the crew:

> Investigators were able to identify a series of errors that initiated with the flight crew's acceptance of the controller's offer to land on runway 19 ... The CVR indicates that the decision to accept the offer to land on runway 19 was made jointly by the captain and the first officer in a 4-second exchange that began at 2136:38. The captain asked: "would you like to shoot the one nine straight in?" The first officer responded,

"Yeah, we'll have to scramble to get down. We can do it." This inter-change followed an earlier discussion in which the captain indicated to the first officer his desire to hurry the arrival into Cali, following the delay on departure from Miami, in an apparent effort to minimize the effect of the delay on the flight attendants' rest requirements. For example, at 2126:01, he asked the first officer to "keep the speed up in the descent" ... The evidence of the hurried nature of the tasks performed and the inadequate review of critical information between the time of the flightcrew's acceptance of the offer to land on runway 19 and the flight's crossing the initial approach fix, ULQ, indicates that insufficient time was available to fully or effectively carry out these actions. Consequently, several necessary steps were performed improp-erly or not at all.

(Aeronautica Civil 1996: 29)

It is easy to pick through the voice record of an accident sequence and find fragments that all seem to point to a common condition. The investi-gator treats the voice record as if it were a public quarry to select stones from, with the accident explanation being the building he needs to erect. The problem is that each fragment is meaningless outside the context that produced it: each fragment has its own story, background, and reasons, and at the time it was produced it may have had nothing to do with the other fragments that it is now grouped with. Also, various behaviors take place in between the fragments. These intermediary episodes contain changes and evolutions in perceptions and assessments that separate the excised fragments not only in time, but also in meaning. Thus, the condition that binds prima facie similar performance fragments arises not from the circumstances that brought each of the fragments forth; it is not a feature of those circumstances. Instead, it is an artifact of the investigator. In the case described above for example, "hurry" is a condition identified in hindsight, one that plausibly couples the start of the flight (almost two hours behind schedule) with its fatal ending (on a mountainside rather than in an airport). "Hurry" is a retro-spectively invoked leitmotiv that guides the search for evidence about itself. This leaves an investigation not with findings but with tautologies.

Aircraft accident analysis as social reconstructions: four characteristics

The first characteristic of aircraft accident analysis is that it is obviously retro-spective, which has importance for the conclusions that are made. While investigations aim to explain the past, they are conducted in the present and thus inevitably influenced by it. One safe assumption is that investigators know more about the mishap in question than the people who were caught up in it. Investigators and other retrospective observers know the true nature of the situation surrounding the people at the time (that is, where they were versus where they thought they were, what mode their system was in versus what mode they thought it was in, and so forth). Investigators also know the outcome of the situation and inevitably evaluate people's assessments and actions in the light of this outcome. The effects of this hindsight bias have been well documented and even reproduced under controlled circumstances (Fischoff 1975). One effect is that "people who know the outcome of a complex prior history of tangled, indeterminate events remember that history as being much more determinant, leading 'inevitably' to the outcome they already knew" (Weick 1995: 28). Hindsight allows us to change past indeterminacy and complexity into order, structure, and oversimplified causality (Reason 1990). The problem is that *structure is itself an artifact of hindsight* – only hindsight can turn ill-structured, confusing and blurred events into clear decision paths with junctures that show where practitioners went the wrong way. Another problem is that, to explain failure, we seek failure. We seek the incorrect actions, the flawed analyses, and the inaccurate perceptions, even if these were not thought to be influential or obvious at the time (Starbuck and Milliken 1988). This search for people's failures is the result of the hindsight bias by which knowledge of outcome influences how we see a process. If we know the outcome was bad, we can no longer objectively look at the behavior leading up to it without concluding that it must also have been bad (Fischoff 1975; Woods *et al.* 1994).

Aircraft accident analysis as proximal, counterfactual and judgmental

It is not just retrospective knowledge of outcome and wider circumstances that colors the interpretation of data on past behavior. Investigations are governed by implicit goals that go beyond merely "understanding" what went wrong and preventing recurrence. Mishaps are surprising in relation to prevailing beliefs and assumptions about the system in which they happen (Wagenaar and Groeneweg 1987). Thus investigations are inevitably affected by a concern to reconcile a disruptive event with existing views and beliefs about the system. In this process of reconciliation, something has to "give way" – that is, either the event or the existing beliefs about the system.

In the immediate aftermath of failure, people and organizations may be willing to question their underlying assumptions about the system they use or operate. Perhaps things are not as safe as they previously thought, or perhaps the system contains error-producing conditions that could have spawned this kind of failure earlier (or worse, could do it again). Typically this openness does not, however, last long. As the shock of an accident subsides, parts of the system mobilize to contain systemic self-doubt and change the fundamental surprise into a mere local hiccup that temporarily ruffled an otherwise smooth operation. The reassurance is that the system is basically safe and that it is only some people or some parts in it that are unreliable. In the end, it is not often that an existing view of a system "gives in" to the reality of failure. Instead the event, or the players in it, are changed to fit existing assumptions and beliefs about the system – rather than the other way around. Expensive lessons about the system as a whole and the subtle vulnerabilities it contains can go completely unlearned (e.g. Rasmussen and Batstone 1989; Wilkinson 1994).

Human factors investigations in aviation must be understood against the backdrop of the "fundamental surprise error" (Lanir 1986). The inability to deal with the fundamental surprise of a failure is evident in investigations that are:

- proximal, that is, they concentrate on local perpetrators (those practitioners who are closest to "causing" and potentially preventing the failure) who are seen as the sole engine of action;
- counterfactual, that is, they state what these practitioners could have done to prevent the outcome;
- judgmental, that is, they state how practitioners "failed" to take these actions or notice data that they should have noticed.

An investigation's emphasis on the *proximal* ensures that the mishap remains the result of a few uncharacteristically ill-performing individuals who are not representative of the system or the larger population of practitioners. It leaves existing beliefs about the basic safety of the system intact. *Counterfactuals* enumerate the possible pathways that practitioners could have taken to prevent the mishap (and that look so obvious in hindsight). Much investigative energy is invested in proving what could have happened if certain minute or utopian conditions had been met, and the result of these efforts is often accepted as explanatory. Counterfactuals may be fruitful in exploring potential countermeasures against the same kind of failure, but they essentially explain nothing: describing what did not happen does not explain what happened. Counterfactuals also have no empirical justification, as there are no events in the world that beg their kind of illumination.

The step from counterfactual to *judgmental* in aircraft accident analysis is a small one that is often taken by investigators. Judgments occur when counterfactual pathways are held up as the ones that *should* have been taken, and they slip into investigations surprisingly easily. Instead of explaining events from the point of view of the pilots that are experiencing them, the investigation lays out counterfactual pathways and emphasizes that these pathways should have been taken – which is obvious from hindsight – and asserting that not doing so constituted a failure. None of this reasoning explains the practitioners' actual behavior. Counterfactual and judgmental language is typical of aircraft accident investigations and justifies their proximal emphasis. The message is that if only these people had done something different (which was so obvious! How could they have missed it!), then the accident would not have happened. The reason for failure has been located. Judgments truncate the need for deeper probing into the systemic conditions that perhaps laid the groundwork for controversial proximal assessments and actions in the first place.

Lay accounts

As indicated earlier, another set of reasons for the desituating of data in aircraft accident analysis is the paucity of agreed-upon methods for matching the behavioral particulars of an accident sequence with articulated, theory-driven models of human performance. In this sense, human performance data analysis lags behind the investigation of, say, structural crash evidence where recovered bits and pieces can be contrasted against quantitative models of component performance or tests of identical components under similar circumstances (McIntyre 1994). Conclusions and causal statements flowing forth from such investigations have a measure of reliability – that is, they could in principle be replicated. In contrast, human factors investigators are left to draw inferences and produce ad hoc assertions that bear some relationship with an ill-defined psychological or sociological phenomenon. For example:

- "Deficient situation awareness is evident" (Aeronautica Civil 1996: 34).
- "The CRM (Crew Resource Management) of the crew was deficient" (Aeronautica Civil 1996: 47).
- "They had lost situation awareness and effective CRM" (Aeronautica Civil 1996: 48).

Such "explanations" frequently become conclusions and statements of cause. For example, "Aeronautica Civil determines that the probable causes of this accident were: '. . . (3) the lack of situational awareness of the flightcrew . . .'" (1996: 57). Another factor that contributes to the problem is the fact that the human performance models that investigators use or have available for use (for example, "loss of situation awareness or loss of CRM") are often theory-begging lay interpretations that do little more than parrot popular contemporary consensus between experts and non-experts on the nature of an everyday phenomenon, for example getting lost or confused (Hollnagel 1998). In response to the increasing dominance of human factors in aviation accidents over the past decades, investigators (as well as researchers) have introduced, borrowed or overgeneralized concepts that attempt to capture critical features of individual or coordinative crew behavior (for example, features such as workload, complacency, stress, and situation awareness).

While easily mistaken for deeper insight into psychological factors, these concepts can in fact create more confusion than clarity. A credible and detailed mapping between the context-dependent (and measurable) particulars of an accident sequence and pertinent conceptual models is often lacking. Lay accounts reflect instead inexplicit connections between the behavioral particulars being measured and the condition they indicate. Lay interpretations typically lack the human performance measures or probes that would be necessary to reach down to the context-specific details because these interpretations postulate no underlying psychological theory that could deliver any such measures (Hollnagel 1998). For example, when one pilot asks the other "where are we?" (Aeronautica Civil 1996: 33), to a lay observer this question may be a clear instance of a loss of situation awareness. But there is nothing inherent to models of situation awareness (see, for example, Endsley 1999) that dictates that this would be so. Such models do not propose any performance measurement based on an underlying psychological theory – for example, that asking a question involving direction indicates a loss of situation awareness. This issue is similar to the one faced by psychological field studies, where the translation from context to generalized concept would really need to go through various steps or levels in order for those without access to the original situation to trace and appreciate the conceptual description (Hollnagel *et al.* 1981; Woods 1993). Through lay accounts, however, the jump from accident details to conceptual conclusions is typically a single and large one, supported by the kind of cherry-picking described above – which immunizes it against critique or verification.

Lay accounts afford no (and require no) analysis between context and concept because the link is obvious. In lay accounts, overlap between the model and the modeled is assumed to speak for itself, since popular belief dictates that this is so. Through observation that is guided by lay interpretation, groups of phenomena can be lumped together as one and called a "fact" (through cherry-picking). The lay account subsequently insists on a fundamentally untestable (and thus unfalsifiable) mapping between the phenomenon or quantity being observed (for example, various remarks scattered across the voice record) and the quality under investigation (for example, hurry).

To improve this situation, social science can offer theory-based, well-articulated conceptual models of human performance. However, similar to the issue that confronts field experimentation (Hoffman and Woods 2000), problems occur because there is little guidance to help investigators establish the mapping between on the one hand, the particulars of context-bound behavior in an accident and on the other hand, models or descriptions of human performance that offer a sufficient level of conceptual detail for those particular kinds of settings (see Woods 1993 and Klein 1998 for some achievement in this respect). The lack of guidance is surprising in an era in which human factors are judged to be the dominant contributor to system failure (see, for example, NTSB 1994; Boeing 1994).

Conclusion: resituating the data

The fundamental surprise error means that large, complex events get picked apart and reinscribed to fit existing assumptions about the system in which they took place – rather than that our assumptions get changed so that they can accommodate the possibility of failure. Aircraft accident analysis contributes to this error through retrospective desituating of data, proximal focus, and counterfactual and judgmental language. As a result, different and mutually exclusive investigations can emerge, each emphasizing and relying on its own sets of "cherries" as derived from a single sequence of events (see again Table 3.1). The result unfortunately leaves the larger airline system or industry none the wiser.

How can we prevent the desituating of data? In order to understand the actual meaning that data had at the particular time and place it was produced, investigators need to step into the past themselves (Vaughan 1996). According to Tuchman:

> Every scripture is entitled to be read in the light of the circumstances that brought it forth. To understand the choices open to people of another time, one must limit oneself to what they knew; see the past in its own clothes, as it were, not in ours.
>
> (1981: 75)

When analyzed from the perspective of the context that produced and surrounded it, human behavior is inherently meaningful. The challenge for an investigator is to reconstruct the unfolding situation as it looked to the people whose controversial actions and assessments are under investigation. Taking the perspective of people inside the situation means adopting the local rationality principle: that is, these people were doing reasonable things given their point of view, their available knowledge, their objectives and their limited resources (De Keyser and Woods 1990; Woods *et al.* 1994).

The local rationality principle implies that investigations should not try to find where people went wrong but rather try to understand how people's assessments and actions made sense at the time, given the circumstances that surrounded them. This requires the investigator to reconstruct a picture of the following phenomena:

- how a process and circumstances unfolded around people;
- how people's assessments and actions evolved in parallel with their changing situation;
- how features of people's tools, tasks, and organizational and operational environment influenced their assessments and actions.

In other words, the reconstruction of a practitioner's unfolding mindset begins not with the mind. Instead, it begins with the situation in which that mind found itself. Note how this confirms the basic findings about situated cognition: if we understand the circumstances in which human cognition took place, we will begin to understand those cognitive activities. Such reconstruction also diminishes an investigator's dependence on broad psychological (lay) labels, instead creating a more direct coupling between (possible) perceptions and actions.

References

Aeronautica Civil (1996) *Aircraft Accident Report: Controlled Flight into Terrain, American Airlines Flight 965, Boeing 757–223, N651AA near Cali, Colombia, December 20, 1995*, Bogotá, Colombia: Aeronautica Civil.

Baker, S. (1999) "Aviation Incident and Accident Investigation," in D.J. Garland, J.A. Wise and V.D. Hopkin (eds) *Handbook of Aviation Human Factors*, Mahwah, NJ: Erlbaum.

Billings, C.E. (1997) *Aviation Automation: The Search for a Human-centered Approach*, Mahwah, NJ: Erlbaum.

Boeing Commercial Airplane Group (1994) *Statistical Summary of Commercial Jet Aircraft Accidents: Worldwide Operations 1959–1993* (Boeing Airplane Safety Engineering B-210B), Seattle, WA: Boeing.

—— (1996) *Boeing Submission to the American Airlines Flight 965 Accident Investigation Board*, Seattle, WA: Boeing.

Clark, A. (1997) *Being There: Putting Brain, Body and World Together Again*, Cambridge, MA: MIT Press.

Degani, A. and Wiener, E.L. (1991) "Philosophy, Policies and Procedures: The Three P's of Flightdeck Operations," paper presented at the Sixth International Symposium on Aviation Psychology, Columbus, OH, April 1991.

De Keyser, V. and Woods, D.D. (1990) "Fixation Errors: Failures to Revise Situation Assessment in Dynamic and Risky Systems," in A.G. Colombo and A. Saiz de Bustamante (eds) *System Reliability Assessment*, The Netherlands: Kluwer Academic.

Dekker, S.W.A. and Hollnagel, E. (eds) (1999) *Coping with Computers in the Cockpit*, Aldershot, UK: Ashgate.

Endsley, M.R. (1999) "Situation Awareness in Aviation Systems," in D.J. Garland, J.A. Wise and V.D. Hopkin (eds) *Handbook of Aviation Human Factors*, Mahwah, NJ: Erlbaum.

Fischoff, B. (1975) "Hindsight is not Foresight: The Effect of Outcome Knowledge on Judgement under Uncertainty," *Journal of Experimental Psychology: Human Perception and Performance* 1(3): 288–299.

Flight Safety Foundation (1997) "Preparing for a Last-minute Runway Change, Boeing 757 Flight Crew Loses Situational Awareness, Resulting in Collision with Terrain," *FSF Accident Prevention* July–August: 1–23.

Gibson, J.J. (1979) *The Ecological Approach to Visual Perception*, Boston, MA: Houghton-Mifflin.

Hoffman, R.R. and Woods, D.D. (2000) "Studying Cognitive Systems in Context," *Human Factors* 42(1): 1–7.

Hollnagel, E. (1998) "Measurements and Models, Models and Measurements: You Can't Have One Without the Other," *Proceedings of NATO AGARD Conference*, Edinburgh, UK.

Hollnagel, E., Pederson, O.M. and Rasmussen, J. (1981) *Notes on Human Performance Analysis (Tech. Rep. Riso-M-2285)*, Denmark: Riso National Laboratory.

Hutchins, E. (1995) *Cognition in the Wild*, Cambridge, MA: MIT Press.

Klein, G. (1998) *Sources of Power: How People make Decisions*, Cambridge, MA: MIT Press.

Lanir, Z. (1986) *Fundamental Surprise*, Eugene, OR: Decision Research.

McIntyre, J.A. (1994) "Perspectives on Human Factors: The ISASI Perspective," *Forum* 27(3): 18.

National Transportation Safety Board (1994) *Safety Study: A Review of Flightcrew-involved Major Accidents of U.S. Air Carriers, 1978 through 1990* (NTSB/SS–94/01). Washington, DC: NTSB.

Neisser, U. (1976) *Cognition and Reality*, San Francisco, CA: Freeman.

Orasanu, J. and Connolly, T. (1993) "The Reinvention of Decision Making," in G.A. Klein, J. Orasanu, R. Calderwood and C.E. Zsambok (eds) *Decision Making in Action: Models and Methods*, Norwood, NJ: Ablex.

Rasmussen, J. and Batstone, R. (1989) *Why do Complex Organizational Systems fail?*, Environment Working Paper No. 20. Washington, DC: World Bank.

Reason, J. (1990) *Human Error*, Cambridge, UK: Cambridge University Press.

—— (1997) *Managing the Risks of Organizational Accidents*, Aldershot, UK: Ashgate.

Starbuck, W.H. and Milliken, F.J. (1988) "Challenger: Fine-tuning the Odds Until Something Breaks," *Journal of Management Studies* 25: 319–340.

Suchman, L.A. (1987) *Plans and Situated Actions: The Problem of Human–Machine Communication*, Cambridge, UK: Cambridge University Press.

Tuchman, B.W. (1981) *Practicing History: Selected Essays*, New York, NY: Norton.

Varela, F.J., Thompson, E. and Rosch, E. (1995) *The Embodied Mind: Cognitive Science and Human Experience*, Cambridge, MA: MIT Press.

Vaughan, D. (1996) *The Challenger Launch Decision*, Chicago, IL: University of Chicago Press.

Wagenaar, W.A. and Groeneweg, J. (1987) "Accidents at Sea: Multiple Causes and Impossible Consequences," *International Journal of Man–Machine Studies* 27: 587–589.

Weick, K. (1995) *Sensemaking in Organizations*, London: Sage.

Wilkinson, S. (1994) "The Oscar November Incident," *Air & Space*, February–March.

Willems, E.P. and Raush, H.L. (eds) (1969) *Naturalistic Viewpoints in Psychological Research*, New York, NY: Holt, Rinehart and Winston.

Winograd, T. (1987) *Three Responses to Situation Theory (Technical Report CSLI-87–106)*, Stanford, CA: Center for the Study of Language and Information, Stanford University.

Winograd, T. and Flores, F. (1987) *Understanding Computers and Cognition*, Reading, MA: Addison-Wesley.

Woods, D.D. (1993) "Process-tracing Methods for the Study of Cognition Outside of the Experimental Laboratory," in G.A. Klein, J. Orasanu, R. Calderwood and C.E. Zsambok (eds) *Decision Making in Action: Models and Methods*, Norwood, NJ: Ablex.

Woods, D.D., Johannesen, L.J., Cook, R.I. and Sarter, N.B. (1994) *Behind Human Error: Cognitive Systems, Computers and Hindsight*, Dayton, OH: CSERIAC.

Part II

Defining risks

Introductory comments

This part focuses on the interactive processes by which groups of actors interpret and negotiate risks in sociotechnical systems. The chapters are grounded in case studies of two systems that are highly contested in many Western societies, namely automobility and systems for handling nuclear waste. The authors analyze the heterogeneous ways in which various actors – such as spokespersons from the automobile industry, government agencies, and environmental groups; workers and managers at a nuclear handling facility, politically appointed inspectors and regulators – attempt to control debates and their outcomes.

Jane Summerton explores the dynamics of a recent controversy in Sweden over the environmental risks of a specific type of automobile that is increasingly popular in many areas, namely cars with diesel motors. In what ways – and specifically, to what measurable extent – do these cars have fewer harmful emissions than conventional gasoline cars? This issue is hotly contested due to the high political and economic stakes for powerful actors (particularly automobile manufacturers and regulators) in both Sweden and continental Europe. The chapter analyzes the intertwining of texts, procedures and tests, negotiations and politics through which risks are defined and contested in actual practices within a variety of forums. The participants in these interactions are typically experts in their areas, and the controversy is played out in many expert forums in which the public – as well as less powerful groups who lack access to these forums – are implicitly excluded.

Andrew Koehler discusses the politics of exclusion in an even more striking sense, namely as an explicit strategy on the part of workers and managers at a plutonium handling facility to avoid having to explain and defend their everyday procedures and practices. By "bunkering" within their organization, these operators seek to construct and maintain strict boundaries between themselves and those actors who might present challenges to their ways of thinking, their common understandings and their ways of handling the hazards of their work. Koehler shows, however, that this practice ironically has high costs and high stakes. Because they do not develop strong ties to the public, regulators or policy makers, the operators are not able to influence political processes as effectively as might be needed to protect the interests of the facility. The result is mutual distrust among actors, as well as a defensive operational stance that, in turn, has a potentially negative outcome for the ability to maintain satisfactory risk handling and safety.

What are some of the interesting conclusions to emerge from these chapters?

First, the case studies vividly demonstrate an important point of departure for this book, namely that stabilized interpretations of risk are *collective achievements* in which power relations, conflicting perspectives and differences in knowledge among actors shape the negotiated meanings that emerge. This constant intertwining of artifacts, procedures, practices and politics affirms the social nature of sense making about the risks of complex sociotechnical systems.

Second, the studies illustrate the ways in which *risk and politics are continually co-constituted* in processes of interpreting and negotiating technological risks. The issue of "who gets to decide" the meaning of a risk and its implications is a contested issue with strong political implications. Powerful groups of actors use constructions of risk as resources to maintain those dimensions of the social and political order that affirm and sustain their positions. Thus spokespersons for the powerful automobile industry in Sweden use their interpretations of specific emission levels of diesel cars as political weapons in ongoing contestations with transnational regulators. Similarly, workers and managers in a plutonium waste facility use "bunkering" as a political strategy (albeit with unintended results) in their conscious attempts to limit public engagement and avoid undue regulatory attention.

The chapters in Part II point to an important theme that is relatively underexplored in existing social science approaches to risk, namely the ways in which risks and politics are co-constituted and reconstituted in everyday technological practices. What are the mechanisms by which politics and power relations are asserted and intertwined in negotiations and contestations over "whose risk" should be accepted, legitimized and stabilized as a "fact"? How might groups of non-represented, excluded or marginalized actors – such as the environmental groups who have radically different views of the risks of automobility than those of the car industry, or the rural communities in close geographic proximity to high risk facilities – resist, intervene or challenge the hegemonic "prevailing views" of more powerful actors? And what might be the political implications of these contestations?

Finally, it is clear that work within the sociology of risk that focuses on everyday practice in organizations and sociotechnical systems needs to be informed by recent work on communities of practice (Chaiklin and Lave 1993; Wenger 1998) and situated knowledges and situated action in specific settings (Suchman 1987; Lave and Wenger 1991; Berner 2000). An interesting example of the fruitfulness of applying these approaches to studies of risk constructions is provided by Gherardi, whose work combines an analysis of communities of practice with perspectives on the situated learning of safety in a discussion of the social construction of safety within organizations (Gherardi 2000; Gherardi and Nicolini 2000).

References

Berner, Boel (2000) "Manoeuvring in Uncertainty: On Agency, Strategies and Negotiations," in B. Berner and P. Trulsson (eds) *Manoeuvering in an Environment of Uncertainty. Practices and Structures in Sub-Saharan Africa*, London: Ashgate Publishers.

Chaiklin, S. and Lave, J. (1993) *Understanding Practice: Perspectives on Activity and Context*, Cambridge UK: Cambridge University Press.

Gherardi, S. (2000) "The Organizational Learning of Safety in Communities of Practice," *Journal of Management Inquiry* 9(1): 7–19.

Gherardi, S. and Nicolini, D. (2000) "To Transfer is to Transform: The Circulation of Safety Knowledge," *Organization* 7(2): 329–348.

Lave, J. and Wenger, E. (1991) *Situated Learning: Legitimate Peripheral Participation*, Cambridge, UK: Cambridge University Press.

Suchman, L. (1987) *Plans and Situated Actions*, Cambridge, UK: Cambridge University Press.

Wenger, E. (1998) *Communities of Practice: Learning, Meaning, and Identity*, Cambridge, UK: Cambridge University Press.

4 "Talking diesel"

Negotiating and contesting environmental risks of automobility in Sweden

Jane Summerton

The environmental risks of automobiles have been widely acknowledged in most Western countries for several decades. In Sweden, as in many other countries, an awareness of the harmful emissions from passenger cars emerged in the 1960s. Since that time, the contested issues of relative levels of emissions, their environmental implications and how these implications should be interpreted have been debated in relation to numerous specific controversies surrounding the automobile and its role. While some of these debates have been markedly public (such as controversies over new highways, bridges or other infrastructural projects), other less "visible" discussions are part of the everyday, business-as-usual practices by which various groups of actors carry out their ongoing work of interpreting and negotiating risks and their social meanings.

This chapter explores negotiations, conflicts and politics among groups of actors that make different claims about the environmental risks of one artifact within the extensive sociotechnical system of automobility. Specifically, the chapter will focus on recent discussions in Sweden over emissions of carbon dioxide (CO_2) from an increasingly popular type of car, namely cars with diesel engines. The environmental risks of diesel cars have been a source of ongoing debate and controversy among groups of actors from government agencies, the automobile industry and environmental groups for several years. Why are these dynamics interesting from a constructivist perspective on risk? First, the recent Swedish debate on CO_2 emissions from diesel cars provides an opportunity to analyze the discursive strategies that actors use to advance their own claims and contest and destabilize those of others. Second, as a site of conflicting commitments and interpretations with high political stakes, the debate reflects broader issues concerning the environmental risks of automobility, the interests and politics that surround these risks, as well as the ways in which technology, politics and constructions of risk are closely intertwined in struggles over the future of the car.

The purpose of the chapter is thus to analyze (1) Swedish actors' constructions of the CO_2 emissions of diesel cars as these constructions have evolved in negotiations and conflicts among these actors, and (2) actors' discursive strategies for putting forth their claims and contesting the claims of others.

The chapter will "follow the claims" of specific groups of actors through selected sites of explicit negotiation over the basic point of contention, namely the issue of CO_2 emissions from diesel cars in relation to emissions from gasoline-driven cars. The chapter is organized as follows. First, I will discuss the theoretical perspectives on discursive strategies for constructing environmental risks that provide the framework for the discussion. The empirical study follows, starting with a brief explanation of what character- izes a diesel motor as a specific type of automobile technology. The chapter will then trace the multiple ways in which various actors interpret, nego- tiate and contest the environmental risks of diesel cars, focusing on the CO_2 issue. What groups of actors are engaged in the controversy and what is at stake for these actors? What negotiations and conflicts have taken place and at what sites, that is, where do the various groups of actors "meet" and interact, and how have these interactions shaped the content and outcome of the controversy? What constitutes the broad politics within which the discourse is embedded and shaped? The chapter will conclude by summar- izing five discursive strategies by which various actors attempt to stabilize and/or strengthen their claims while destabilizing those of others, asking what these strategies tell us about the politics of risk constructions.

Constructing claims, contesting others' claims

Constructivist studies of risk focus on the processes by which risks are constructed, negotiated and often contested by heterogeneous actors or groups of actors (see, e.g. Bradbury 1989; Nelkin 1989; Wynne 1989; Hilgartner 1992; Hannigan 1995; Schmid 2001). A basic premise of the constructivist approach is that multiple social, cultural and institutional commitments are "embedded and reproduced, usually tacitly, within technical risk assessments that typically purport to be objective" (Zwanenberg and Millstone 2000: 259). The varying assessments thus reflect and embody the involved actors' interests, values and commitments with regard to the technology at hand and its risks. In debates or controversies over technological risks, actors engage in a series of negotiations in their interactions with other groups of actors, each of whom competes to define, stabilize and control the emergent cultural meanings of risk in relation to the technology.

Controversies over technological risks thus often entail discursive strug- gles (Lidskog *et al.* 1997; Schmid 2001) in which competing claims are put forth, contested and defended among the various groups of actors. While some actors will attempt to frame the issues in ways that advance their particular commitments, values or interests, others may try to challenge and destabilize these constructions on the basis of their own agendas. Actors may seek to control not only the immediate discourse at hand but also the dynamics of power in broader political processes or struggles that have signif- icantly high stakes. In a study of discursive shifts in the representation of nuclear risks in post-Soviet media and politics, Schmid, for example, notes:

how the discourses about risks and uncertainties of nuclear energy are themselves used as technologies, as rhetorical tools strategically aimed at gaining, consolidating, or protecting control over decision-making processes in the energy sector, and, as a result, in policy making in general.

(Schmid 2001: 2)

Controversies about risk are thus often *sites of political contestation and struggle* where "risk, politics and culture are simultaneously co-produced" (Parthasarathy 2001: 3). The ultimate stabilizing of specific interpretations of risk is a political achievement, and when networks of powerful actors succeed in shaping the discourse, they also influence the ways in which the cultural meanings of the technology are interpreted.

What are some of the discursive strategies that actors use to advance their own claims about risks, destabilize the claims of others and shape debates over technology and their outcomes? There are at least three types of strategies that are suggested by the relevant literature about discourses on environmental risks and safety. First, one discursive strategy is to *displace others' risk objects* (Hilgartner 1992). Hilgartner notes that definitions of risk encompass three conceptual components: a risk object that is viewed as the source of risk, a harm that is attributed to this object and a linkage that is viewed as causally connecting the risk object and harm. Actors will often attempt to displace others' risk objects by deconstructing other actors' definitions of risk objects and/or severing their linkages to harm, thereafter replacing them by constructing new objects as sources of risks to be reckoned with. In a study of the construction of risks in automobility, Hilgartner shows how consumer advocate Ralph Nader successfully challenged "the driver" as the dominant risk object in debates about road safety in the 1960s, replacing it with a new network of risk objects that included the engineers who designed unsafe cars, the managers who cut production costs, the regulators who failed to supervise automotive design and the politicians who did not legislate such supervision (Hilgartner 1992: 46–50). Hannigan (1995) also provides numerous examples of risk debates in which the definition of harm as linked to a risk object is contested, resulting in multiple claims and counterclaims even when there is agreement as to the risk object.

A second discursive strategy by which actors seek to advance their own claims and contest those of other actors is *discursively transforming calculable risks into non-calculable uncertainties* (Schmid 2001: 15). Schmid shows, for example, how anti-nuclear activists deconstructed the certainty of Soviet technocratic statements in the aftermath of the Chernobyl accident. While technocratic proponents of nuclear power emphasized risks (as reflected in design flaws that were known in advance and calculable), the activist opponents emphasized the uncertainty of the human factor as fundamentally incalculable (Schmid 2001: 15–16). The opponents continually challenged many aspects related to nuclear power and the nuclear industry, underscoring the multiple uncertainties that were involved, including uncertainties in scientific knowledge. In a similar argument, Dratwa discusses the use of the precautionary principle in environmental discourses of the 1990s as a means "precisely to open up risk analysis and policy making to situations of uncertainty, situations which are controversial scientifically as well as societally" (Dratwa 2001: 10).

Related to these dynamics, a third discursive strategy in environmental discourses about risk is to *challenge the procedures of science* by questioning the validity, appropriateness or local compatibility of these procedures. An example is provided by Wynne's (1989) study of communication among government experts and Cumbrian sheep farmers over the extent to which local grazing areas, and in particular various kinds of soils, were contaminated by radioactive fallout from Chernobyl. Wynne illustrates how the farmers and their advocates were ultimately able to challenge both the expert knowledge of the government chemists and the local (lack of) compatibility of the scientific tests upon which this knowledge was based.

In analyzing the processes by which various actors recently negotiated and contested the environmental risks of diesel cars in Sweden, this chapter will focus on interactions among spokespersons for three groups of actors, namely

1   government agencies (specifically the Swedish Environmental Protection Agency and the Swedish National Road Administration);
2   Swedish car manufacturers and their interest groups (primarily Volvo Cars and BIL Sweden); and
3   environmental groups (specifically the Swedish Society for Nature Conservation, SNF, and Green Motorists).

These organizations have been chosen because they have been active in the recent Swedish debate on diesel cars and their environmental risks, particularly the issues of CO_2 emissions and particle emissions from this type of car. The actors have been identified through analysis of various written materials as well as through interviews with various actors in the debate. The empirical data for the study consists of different types of written materials (for example, government studies, press articles, materials and documents from the various organizations, reports, websites), interviews with spokespersons from the involved organizations,[1] and participant observation in conferences in 1999 and 2000.

The "black box" under the hood: what is a diesel motor and what is the fuss about?

The diesel motor was first conceptualized around 1890 by Rudolf Diesel, whose ambition was to create an engine that operated as close to the theoretically maximal degree of efficiency as possible (that is, according to the Carnot process, so called after the French engineer Sadi Carnot's calculations from the 1920s). A diesel motor is a combustion motor that self-ignites when diesel fuel is injected into highly compressed air in a cylinder. The diesel is able to self-ignite because the temperature of the compressed air exceeds the temperature at which the injected fuel burns, and the fuel ignites shortly after it enters the cylinder. Two main types of diesel motors can be distinguished, namely the direct injection (DI) motor and the indirect injection (IDI) motor. In the case of the former, the fuel is injected directly into the cylinder chamber, while in the latter, fuel is injected into an adjoining chamber that is connected with the cylinder by a narrow channel. IDI motors are usually used in cars due to their compactness and higher fuel efficiency in relation to motor capacity.

Diesel cars have a higher energy efficiency compared with cars that have conventional gasoline-driven motors. Although assessments vary, some researchers have estimated that the fuel efficiency of diesel cars is about 20–25 percent higher than gasoline cars (see, for example, Lenner 1998: 41). Because they use less fuel, diesel cars also emit relatively lower amounts of carbon monoxide (CO) in comparison with gasoline-driven cars. On the other hand, diesel cars emit higher levels of nitric oxides (NOx) and – in particular – *soot particles,* which have been identified as a significant risk in recent years (Swedish National Road Administration 1999a: 5) but will not be further discussed here.

The issue of carbon dioxide (CO_2) emissions from diesel cars has, however, been described as "the absolute most difficult one"[2] to deal with. Because of the continual oversupply of air in diesel motors, the amount of CO_2 that is emitted from such cars is lower than CO_2 emissions from comparable gasoline-driven cars. But as this chapter will show, the crucial question is *how much lower*; that is, to what extent are the CO_2 emissions of diesel cars lower than the emissions of comparable gasoline-driven cars? Are diesel cars environmentally preferable to gasoline-driven cars, and if so, according to whom, by what testing procedures and by what interpretations? Based on these interpretations, should environmentally aware consumers be advised to purchase diesel cars?

In many parts of Europe, passenger cars that are equipped with diesel motors have been popular among motorists for many years. For example, in France 40 percent of all registered cars reportedly have diesel motors, while in Austria over 50 percent have such motors (BIL Sweden 1999c: 2). In contrast, Sweden has until recently had a low percentage of diesel-driven cars on its roads. In the 1990s, however, diesel cars increased significantly in popularity among Swedish car owners. The number of diesel cars doubled from 1996 to 1998 alone, and the trend is expected to continue. By 2000, nearly one in six new car registrations had diesel technology. The increase is due to several developments, most notably a marked increase in imports of diesel cars since Sweden's entry into the European Union, changes in domestic taxation which have favored diesel (particularly for motorists with high mileage levels) and a relatively widespread use of diesel cars as business cars.

The unexpected increase in diesel cars has been a cause of concern among Swedish policy makers for both environmental and fiscal reasons. How might increasing numbers of diesel cars influence the chances of reaching environmental goals for road traffic as recently targeted by the Swedish government (see Swedish government proposals 1997/98: 56, 1997/98: 145), and – related to these goals – how should diesel cars be taxed? These questions were the basis for the government's decision in mid-1998 to commission a study on the environmental and health risks of diesel cars.[3] In motivating this study, the government also specifically pointed to an overview of traffic taxation that was currently under way at the time. The task of this overview was to carry out a comprehensive review of Sweden's system of taxation for road traffic, which in turn would form the basis for future decisions about how to balance various types of taxes, including a CO_2 tax, on different types of automobiles. A commissioned study of the environmental and health risks of diesel cars would provide important input for the committee that was carrying out the traffic taxation study.

The "alarm" report by Swedish Environmental Protection Agency, SEPA

In May 1998, the Swedish government commissioned the above-mentioned study on the environmental and health impacts of projected increases in the number of diesel cars on Swedish roads. Responsibility for carrying out the study was given to the Swedish Environmental Protection Agency, SEPA. SEPA is the leading government agency for studying and handling environmental issues in Sweden, operating directly under the Ministry of Environment. Since its establishment in the mid-1960s, SEPA has played a major role in carrying out and coordinating countless environmental studies on behalf of the Swedish Parliament and government.

SEPA's study was carried out by its division for chemical regulations and written by its division chief. Less than six months after it was commissioned, the study was released with the somewhat cumbersome title "What will be the consequences for environment and health of an increased percentage of diesel cars in the Swedish car system?" (see Swedish Environmental Protection Agency 1998).

How does the SEPA report construct the environmental risks of diesel cars in Sweden and, in particular, the politically important risks of CO_2 emissions from such cars? The report differentiates between different types of cars and different types of driving conditions, each constituting a specific variety of "diesel cars." First, the report limits its analysis to emissions of air pollutants "under operating conditions," stating that approaches such as lifecycle analysis are not considered. Second, the report specifically quantifies and compares emissions from four groups of cars that are assumed to have differences with regard to emission levels, namely "new gasoline cars" (i.e. cars in so-called class 1, with modern exhaust cleansing technology); "new diesel cars" (i.e. cars with DI (direct injection) motors and catalytic converters); "old gasoline cars" (i.e. cars from the late 1980s and earlier, without catalytic converters) and "old diesel cars" (i.e. cars from the late 1980s and earlier, without exhaust cleansing technology). For each of these categories, a distinction is also made between "city driving conditions" and "rural driving conditions." Finally, the report puts a price tag on environmental damage, making quantitative estimates of the "environmental cost" of various kinds of emissions by applying concrete assumptions and parameters on how economic costs should be calculated.

Specifically with regard to the CO_2 issue, the SEPA report claims that a new diesel car emits about *8 percent less carbon dioxide* than a new gasoline car. This relatively lower level of emissions has, however, an "insignificant effect on the potential to reach environmental goals for (reducing) climate change." More diesel cars on the road would lead to only "marginal decreases" in emissions of CO_2 (as well as hydrocarbons). At the same time, an increase in the number of diesel car registrations in Sweden from 1 percent to 20 percent of all new registrations – which is viewed as a possible scenario, based on current trends – could be expected to lead to an 80 percent increase in emissions of nitric oxides (NOx) and a huge 240 percent increase in particles. To minimize environmental risks, the results are clear:

1   consumers should purchase an energy-efficient, environmentally marked *gasoline-driven car* rather than a diesel car;
2   policy makers should consider concrete measures to restrict sales of both *new* and *old* diesel cars, particularly imports of old cars, which have increased dramatically in recent years; and
3   the total societal cost of all emissions, expressed as "environmental and health effects per liter of fuel," is about *twice as high* for diesel cars as for gasoline cars.

(Swedish National Road Administration 1998: 52)

SEPA's study is thus unequivocal in its summary findings and recommendations. Viewed in light of current goals for environmental quality that center on alleviating acidification and overfertilization and achieving clean air in cities, the current trend toward more diesel cars in Sweden is going in "the exact opposite direction" than desired (Swedish National Road Administration 1998: 52). The often noted advantage of diesel cars, namely their relatively low CO_2 emissions, can be quantified at *only 8 percent lower* than emissions from gasoline-driven cars.[4] This level in turn has an "insignificant" effect on efforts to reduce anthropogenically induced climate change.

Contesting SEPA's results, negotiating texts

One of the actors who read SEPA's report with considerable interest was a senior officer from the Swedish National Road Administration (VV for

Vägverket). VV is responsible for "promoting an environmentally-adapted road traffic system and an environmentally-adapted vehicle traffic" as a means of achieving environmental goals for road traffic as established by the Swedish Parliament. Its openly ambivalent view of automobility is expressed in the following statement from its home page:

> The car gives us so much in so many different ways. Our material welfare, our work and our leisure time – yes, our whole culture and our society are in many ways built around the car. At the same time, the car and automobility are an increasingly pervading threat to ourselves and our environment. Road traffic as we know it today is not what we call "durably sustainable."
>
> (Swedish National Road Administration 1997: 1; my translation)

The issue of what type of cars – diesel or gasoline driven – that are least detrimental to the environment is a topic that VV has explicitly addressed in numerous written materials throughout the 1990s.[5]

VV's senior officer is critical of SEPA's report on several counts. One point of contestation concerns the *testing procedure* for certifying new car models in Europe, the results of which provided the basis for SEPA's statistics on the relative emissions of diesel and gasoline cars. Accordingly to VV's senior officer, these test procedures reflect considerable uncertainties, as well as a positive bias towards gasoline cars. First, the majority of harmful emissions are released under driving conditions that are not covered by the testing procedure, namely cold starts and rapid accelerations – precisely those situations in which diesel cars have an advantage over gasoline cars. Second, most of the tests in SEPA's study referred to older cars, and there was relatively "little material" to support the study's findings on newer cars. This point is significant, VV's official notes, because of the rapid technological advances in diesel motors that make up-to-date testing essential. Third, according to its standardized procedure, the certification test for new car models only requires the testing of one single car, which means that potential variations are not detected. According to VV's senior officer, a car manufacturer can – at least in theory – "handpick" the car that is to be tested, and then drive it in the most favorable way. The testing procedure is thus not sufficiently tough on both types of cars, asserts VV's senior officer, and its validity can be questioned.[6]

Most important for our discussion here, VV's senior officer maintains that SEPA's report does not sufficiently value the merits of diesel cars in *reducing CO_2 emissions*, compared to gasoline-driven cars. For example, CO_2 emissions from the diesel version of Volkswagen's new Golf model are reportedly 35–40 percent lower than in the gasoline model, according to VV's senior officer.[7]

This claim – that some diesel cars emit *35–40 percent* less CO_2 than comparable gasoline-driven models – stands in sharp contrast to the findings of SEPA's claim that new diesel cars yield only *8 percent* reductions. Soon after the release of SEPA's report, these contrasting claims became a topic of open controversy between the leading experts from SEPA and VV. In fall 1998, experts from VV, SEPA and the Swedish Consumer Authority were in the midst of a cooperative project to update informational materials about their annual "environmental labeling" procedure for cars sold in Sweden. In a series of meetings, one of the materials that was explicitly discussed was VV's 1999 version of its booklet "Safety and environmental impacts of cars." SEPA and VV had cooperated for several years in formulating this booklet, which is intended to provide guidelines for consumers when buying new cars and car dealers when selling them.

Following the release of SEPA's report, however, the exact wording of one part of this booklet became an explicit source of conflict between SEPA's division chief and VV's senior officer, both of whom participated in the meetings at hand. The conflict focused on what information should be provided to readers about precisely the question of to what extent new diesel cars had fewer CO_2 emissions than comparable gasoline cars. In negotiating over the text, VV's senior officer was unwilling to change his initial formulation in the text of "40 percent fewer carbon dioxide emissions" – based on the results of the Volkswagen Golf tests noted above. SEPA's division chief reportedly asserted that VV's "overestimated" the extent to which diesel cars lead to reduced CO_2 emissions. The result was a compromise largely on VV's terms: the words *"up to"* (40 percent) were inserted in the text, as shown in the final published version of the booklet under the heading "Gasoline or Diesel?":

> The most modern diesel cars can . . . release up to 40 percent fewer carbon dioxide emissions compared with the comparable gasoline variety. For the previous generation of diesel cars, carbon dioxide emissions are 10–20 percent lower than for comparable gasoline cars.
> (Swedish National Road Administration 1999b: 9, 1999c: 9–10; my translation)

According to VV's senior officer, the controversy with SEPA over the relative risks of diesel cars continued to be played out in various forums in which the two authorities met, such as projects, seminars and study visits. The senior officer claims that VV's current position is clear: "the gasoline car is better than diesel *health-wise* – diesel is better from a *climate* standpoint" (see, for example, Swedish National Road Administration 1999b: 9). The diesel car should be avoided in *urban areas* because of air congestion; a limited number of diesel cars should be accepted in *rural areas* because of the advantages in reducing greenhouse effects. VV's senior officer also claims that the "problem" of diesel cars is, however, somewhat misdirected: the real problem lies with heavy diesel busses and lorries which emit much more CO_2 than diesel cars. Thus VV wants to *displace* the debate over diesel cars and instead discuss the broader issue of the environmental effects of the entire diesel complex.

Meanwhile, SEPA's division head remains critical to the text discussed above because it "does not provide good guidance to consumers." He admits, however, that diesel cars' relative contribution to reducing CO_2 levels should be viewed more positively than he originally did in the SEPA report. Instead of 8 percent fewer CO_2 emissions from diesel cars, the most accurate level should be viewed as "about 20 percent fewer CO_2 emissions."[8]

We have seen how SEPA's construction of "8 percent fewer CO_2 emissions" was made to be compatible with VV's "up to 40 percent fewer CO_2 emissions." But how and why did SEPA's original position of "8 percent fewer CO_2 emissions" evolve to the new position of "about 20 percent fewer CO_2 emissions" as just indicated? This new construction of facts evolved in negotiations with experts from Volvo Cars.[9]

Negotiating percentages with the automobile industry

The Swedish National Road Administration, VV, was not alone in its criticism of SEPA's October 1998 report. Another vocal actor in the growing debate was BIL Sweden (BIL), a Swedish branch organization for twenty-five manufacturers and importers of cars, trucks and lorries, including Volvo and Saab. BIL has been active in the diesel debate since the 1960s and regularly publishes statistics and informational materials on various aspects of the country's car parks as part of its ongoing lobbying efforts to influence Swedish policy makers and the general public on issues of importance to its members.

BIL's spokespersons were critical of SEPA's report in ways similar to VV's senior officer: in their view, SEPA's report clearly underestimated the merits of diesel in reducing carbon dioxide emissions. Shortly after SEPA released its report, BIL devoted most of its January 1999 newsletter to discussing the relative merits of diesel cars. Among other things the newsletter explicitly emphasized energy efficiency and the smaller amounts of CO_2 emissions by diesel motors:

> Diesel-driven motors have considerably lower fuel consumption than equivalent gasoline motors, therefore they also emit a smaller amount of carbon dioxide. These are positive characteristics that simply must be utilized in the pursuit of lower energy consumption and reduced emissions of so-called greenhouse gases.
>
> (BIL Sweden 1999a: 1; my translation)

The newsletter also emphasized the importance of new technologies in improving the environmental characteristics of diesel motors (BIL Sweden 1999a: 1–2). BIL's spokespersons were critical of recent research reports that claim that the small particles released by diesel motors have significant negative effects on health. Similar to VV's senior officer, BIL's vice-president asserted that the *testing* in these reports can be criticized because the researchers tested "old technology, old motors – in short, the effect of old technology."[10] To counteract these claims, BIL has considered conducting its own study on the advantages and disadvantages of diesel cars in the hopes of *displacing* SEPA's experts and interpretations with its own.[11]

In detailed written comments to SEPA's report as well as other materials, Volvo Cars' position also reflects a relatively positive view of modern diesel cars: among other things, they have low fuel consumption and "very good emission values" (see Wallman 1998).[12] Both Volvo Cars' expert for mobility and technology assessment and its division chief for development of diesel motors also explicitly claim that Sweden has "Europe's cleanest diesel engines" due to their low sulphur content and other qualities.[13] The emphasis on technological improvements as the key to solving diesel's problems is also expressed in the Volvo Group's Position Statement on diesel from June 1999.[14] Like VV's criticism, this statement also questions the validity of prevailing *testing methods* for measuring particle emissions, which are one of the most hotly debated issues in the current debate (ibid.). Volvo Cars' division chief underscores the *uncertainties* of current knowledge about particle emissions in particular, viewing it as a topic of "scientific dispute."[15] Finally, several Volvo experts claim that the debate on diesel cars represents a focus on the wrong problem: it is illogical to place environmental demands on new diesel cars when it is the *old* ones which cause much more damage, and the real problem lies with heavy lorries and busses which consume 95 percent of all diesel oil in Sweden.[16]

Meanwhile, in November 1998, a few weeks after the release of SEPA's report, the Swedish Ministry for the Environment initiated an "informal hearing" over the growing controversy over the environmental risks of diesel cars. The hearing, which was held at Volvo's corporate offices in Gothenburg, was attended by SEPA's division chief, various experts from BIL, Volvo and Saab and representatives from the ministry. After a short presentation of SEPA's report, the group of experts discussed the report's most contested findings. Specifically, Volvo Cars' detailed criticism of SEPA's report focused on three main points. First, the findings in the report were not clearly presented but were instead generally difficult to recognize and interpret. The report was written in a negative and opaque way that could lead to "misinterpretations" by laypersons and journalists, who could easily miss many nuances and get an incorrect impression of the results. Second, the report underestimated the positive characteristics of new diesel motors in terms of energy efficiency and fuel consumption (as well as other advantages in relation to old gasoline and diesel cars). Third, SEPA's report underestimated the contribution of diesel cars to reducing CO_2 emissions. On this point, Volvo Cars' expert for Mobility and Technology Assessment asserted at the hearing that SEPA's estimate of 8 percent fewer CO_2 emissions from diesel cars should be corrected to about *20 percent fewer CO_2 emissions* than comparable gasoline-driven cars.[17]

According to initiated sources,[18] the hearing led to a negotiated "agreement" on the following main points: (1) the most pressing problem is the increasing imports of *old* highly polluting diesel cars (which is a finding that is compatible with the automobile industry's interest in increasing sales of *new* cars), (2) SEPA's estimates of the energy efficiency of diesel cars in relation to gasoline cars were too low, and (3) SEPA's estimates of the extent of reduced CO_2 emissions from diesel cars (relative to comparable gasoline cars) were too low.

Following the hearing, SEPA's division chief recalculated his original findings on reduced CO_2 emissions (8 percent) in order to check the validity of Volvo's claim (20 percent), using current data about relative fuel use as indicated in Volvo materials. The results supported, he asserts in an interview, Volvo's Cars' estimate.[19] Because the new calculation did not, however, alter SEPA's main conclusions, the division chief did not revise his report as previously delivered to the government.

SEPA gains allies: the Swedish Society for Nature Conservation (SNF) and Green Motorists

One organization which frequently sends representatives to meetings such as the informal hearing is the Swedish Society for Nature Conservation (SNF), Sweden's oldest and largest environmental organization with 160,000 members. SNF officially supports SEPA's conclusions that new gasoline cars are environmentally preferable to new diesel cars, and that imports of old diesel cars are a significant problem. SNF's expert on traffic issues comments, however, that the organization also supports the interpretation that diesel cars are more acceptable in rural areas than in cities.[20] Even prior to the publication of SEPA's report, SNF had urged readers of its website to purchase a *lean gasoline car* instead of a new diesel car when choosing between gasoline and diesel cars (Swedish Society for Nature Conservation 1995: 1). Similarly, in a 1996 publication, the organization advises against driving a diesel-fueled car (Swedish Society for Nature Conservation 1996: 62).

The issue of the risks of diesel cars is one of the contested topics in which SNF often engages in heated debates, sometimes strengthened by alliances with others who share its positions. One frequent ally is the Swedish Association of Green Motorists. Similarly to SNF, the relatively small Green Motorists explicitly endorses SEPA's conclusions on the environmental effects of diesel cars. The organization specifically advises its members against buying diesel cars, based on their harmful emissions of nitrogen and particles.

Not long after the SEPA report was released, Green Motorists made the following claim concerning diesel cars in its ranking of "the most environmentally benign 1999 cars" in the January 1999 edition of its magazine:

We still maintain that diesel cars should not be used in southern and middle Sweden where the high emissions of nitric oxides contribute to acidification, over-fertilization, and the build up of ozone. Diesel cars also emit 10–15 times more health-damaging particles than comparable gasoline cars. This makes the use of diesel cars totally inappropriate in densely populated areas. In northern Sweden, however, nitric oxides and particles are less of a problem. There the advantage of lower carbon dioxide emissions and better cold-start properties of diesel cars place them on a par with gasoline cars.

> (Green Motorists 1999: 9–10; my translation)

Green Motorists thus claims that the differences in CO_2 emissions from diesel and gasoline cars are highly dependent upon *what types of cars are compared with each other*, specifically what year/model and performance data are chosen, and *what topographies and climates* that these cars are driven in.[21] Notably, in a report for the European Federation for Transport and Environment, a representative for Green Motorists estimates that the difference in *relative emissions of CO_2 was 12 percent less* for 1998 diesel cars than for 1998 gasoline cars with comparable performance data (Kågeson 2000). The report explicitly comments on other actors' claims on the CO_2 issue:

> Spokesmen for the motor industry often say that the difference in CO_2 emissions is around 25 percent, but this is only true for some models with DI (direct injection) "common rail" engines. In other cases comparison may have been made with petrol-fueled variants having power ratings and performance far above those of the diesel versions.
>
> (Kågeson 2000: 17)

Green Motorists also links the CO_2 issue to the amount of driving: diesel cars are probably driven about 15 percent more distance per year than gasoline cars, based on the EU Commission's assumption of long-term fuel price elasticity of −0.7. When this 15 percent factor is compared with the 12 percent lower CO_2 emissions from diesel cars, the result is that the increased use of diesel cars *has not yet resulted in actual reductions of CO_2.*[22]

The CO_2 issue: high political and economic stakes

Why is the issue of relative percentages of CO_2 emissions from diesel cars so fiercely debated by these actors? Why does the diesel dispute matter – and why now?

At a February 2001 public seminar[23] to discuss environmental and health risks of diesel (and other) cars, the head of SNF's Stockholm group claimed that the controversy over the environmental risks of diesel cars must be understood within the context of the so-called EU–ACEA agreement concerning future goals for reducing the CO_2 emissions of automobility.[24] This agreement constitutes an important motivation for legitimizing and promoting diesel motors.

What is the EU–ACEA agreement and why is it so important for Swedish automobile manufacturers? ACEA is an acronym for the European Automobile Manufacturers Association, an important European lobbying group for car manufacturers in Europe in which Volvo, Saab and BIL are members (Tengström *et al.* 1995: 41). One of the objectives of ACEA is to favor "dialogue with the European Community at all levels, and with all others concerned with the automobile industry, including the European public."[25] In July 1998 – just a few months before SEPA released its report – ACEA and the European Commission signed an agreement that contained specific, quantified goals for reducing CO_2 emissions from passenger cars. Specifically, the agreement states that average emissions of CO_2 from all passenger cars that are sold within EU countries *should be reduced by 25 percent* (compared with 1995 levels) by the year 2008.[26]

The importance of diesel cars as a means of reaching this goal is expressed by BIL's chief executive in the organization's January 1999 newsletter (also noted earlier) under the straightforward heading "CO_2 goal demands diesel." The chief executive's comment was specifically intended to be a response to SEPA's report from a few months earlier:

> To "tax away" diesel-driven passenger cars from the Swedish car market would be a big mistake. The diesel motor releases . . . very low emissions of carbon dioxide, CO_2. This makes diesel *totally essential* for the automobile industry to be able to reach 25 percent reduced CO_2 emissions by the year 2008 – the goal that the automobile industry and EU have agreed to in a voluntary agreement.[27]
>
> (emphasis added, my translation)

According to this claim, it is precisely due to their low CO_2 emissions that diesel cars are "totally essential" for current efforts by the automobile industry to achieve the goals of the EU–ACEA agreement.

Diesel cars are thus put forward as part of the *solution* to alleviating existing problems of high CO_2 emissions rather than being a part of the problem of environmental and health risks. Diesel cars are attractive options for car manufacturers for at least two reasons: they are already on the market, and diesel engines can easily be installed in large cars, which car manufacturers tend to favor (and which otherwise have relatively high CO_2 emissions).

The EU–ACEA agreement is also invoked as pressure on Swedish policy makers to avoid policy measures that disfavor diesel cars. For example, Volvo's expert for mobility and technology assessment is cited in BIL's newsletter as explicitly warning that new taxes on diesel cars constitute a direct threat to the EU–ACEA agreement:

> The agreement clearly states that diesel-driven cars are included when the total emissions of carbon dioxide are estimated. This means that the agreement will be *reconsidered* if, for example, Sweden implements

new taxes and rules that hamper introduction of, for example, diesel cars. An increased percentage of diesel is a *pre-requisite* for fulfilling the agreement.[28]

(my translation, emphasis added)

This statement is explicitly reinforced by spokespersons for Swedish Volkswagen and Saab, who are also cited in the newsletter (see BIL Sweden 1999a: 4).

Why are Swedish and other European car manufacturers so firmly committed to the EU–ACEA agreement? After all, the agreement is voluntary, and to reduce CO_2 emissions by a huge 25 percent entails considerable effort and investment. What incentive do car manufacturers have to participate in such a scheme?

The answer is that the EU–ACEA agreement was drawn up with the understanding that *if the manufacturers succeed in reaching this goal, EU will refrain from proposing new environmental legislation (e.g. new taxes, emission norms and standards) on the European auto industry.* As Volvo's expert comments, increased legislation is viewed as a clear economic and political threat, while a voluntary agreement that gives the automobile industry the freedom to solve the problem on its own is a much more attractive option.[29] One important part of the industry's strategy for reaching the CO_2 emissions goal is thus to increase the market share of diesel cars.

In this light, SEPA's report on the detrimental role of diesel cars for health and environment – released just a few months after the EU–ACEA agreement was approved – could hardly have been a welcome turn of events for Volvo Cars, BIL and their European colleagues. While the latter have strong interests in promoting the relative advantages of diesel cars in terms of lower CO_2 emissions, SEPA's report claimed that diesel cars release only 8 percent fewer such emissions than gasoline-driven cars.

Swedish car manufacturers have strong interests in promoting diesel cars for other reasons as well. Volvo Cars is reportedly in the process of developing its own diesel motor for passenger cars as a means of competing on continental European diesel markets; until now, the company has imported Volkswagen's diesel motors for their diesel vehicles.

Continuing contestations . . .

Controversies over the environmental risks of diesel cars continue, although with a shift in focus and among partially new actors. Specifically, the *particle emissions* from these cars have recently been a source of ongoing debate, in which an important new group of actors are *scientists* whose work in measuring and quantifying particle emissions and their health impacts forms the basis for new claims and counterclaims. Along with spokespersons from governmental agencies, car manufacturers and environmental groups, these actors meet in many different fora to "talk diesel."[30]

Conclusion: on malleability, discursive strategies and politics

At least three conclusions can be drawn from the recent Swedish controversy over the environmental risks of diesel cars, particular CO_2 emissions from these cars.

First, the story shows *the malleability of "facts" in the negotiative processes* among actors whereby risks were constructed, interpreted, revised and ultimately stabilized. When SEPA's controversial report was released, the issue of "to what extent do diesel cars release fewer CO_2 emissions than gasoline-driven cars" was a highly contested one in which claims and counterclaims varied greatly:

SEPA – new diesel cars release 8 percent fewer CO_2 emissions than comparable gasoline-driven cars.

VV – new diesel cars release 40 percent fewer CO_2 emissions.

Volvo – new diesel cars release 20 percent fewer CO_2 emissions.

Green Motorists – new diesel cars release 12 percent fewer CO_2 emissions than comparable gasoline cars (but total CO_2 reductions are essentially negligible because diesel cars are driven more).

In subsequent negotiations, these claims proved to be highly malleable. Following the "informal hearing" at Volvo's offices, the author of SEPA recalculated his findings (8 percent), which he now claims supported Volvo's expert's estimate (20 percent). When the experts from SEPA and VV negotiated the wording about CO_2 emissions in their consumer brochure, they compromised on "up to" 40 percent fewer emissions. A report from Green Motorists explicitly refutes "spokesmen for the motor industry" who claim 25 percent fewer emissions, asserting their 12 percent estimate is closer to SEPA's original calculation. Thus as the diesel car travels between different sites of negotiation – texts, hearings, conferences, European tables – "the hard facts" about its environmental risks are alternately contested, refuted, partially withdrawn and modified.

Second, it is clear that actors used *many different types of discursive strategies* to destabilize or contest others' claims while mobilizing support for their own claims. Some of these strategies are recognizable from previous work on the social construction of risk as discussed earlier:

a   Actors worked to *displace* others' definitions of risk, *sever* their linkages to harm and *replace* them with new risk objects (Hilgartner 1992). For example, spokespersons from both the automobile industry and environmental groups attempted to redefine the problem as a broader one (e.g. VV claimed that "the risks of diesel cars are shared by all cars in the future"; SNF claimed that "automobility as a whole is the problem"). Related to this strategy, some actors also attempted to redefine the problem as elsewhere (e.g. VV and Volvo Cars claimed that "the real problem is diesel lorries and other heavy traffic"). Finally, spokespersons from the automobile industry argued for the interpretive flexibility of the linkage between the risk object and environmental harm (e.g. Volvo Cars and BIL claimed that "diesel is a solution – and we can solve its problems" and "Sweden has Europe's cleanest diesel technology").

b   Actors sought to *transform other actors' "risks" into uncertainties* (Dratwa 2001; Schmid 2001) and *destabilize* the certainty of others' constructions. For example, actors underscored the uncertainties in existing scientific knowledge and the lack of scientific consensus on the "risks" at hand (e.g. Volvo Cars and VV claimed that "diesel particles and their effects are a matter of scientific dispute," "we know very little"). Actors also contested and attempted to destabilize the scientific knowledge of others (e.g. Volvo Cars asserted that "SEPA's report is inaccurate on several points").

c   Actors actively *challenged the procedures of science* (Wynne 1989) by questioning the validity of testing procedures (e.g. VV and BIL asserted that "the testing of diesel motors is unreliable").

In addition to confirming these well-known discursive strategies, our discussion has also revealed two other strategies that have not been identified in previous constructivist studies of risk. Specifically, these strategies are:

a   Actors *created bifurcations that deconstructed and destabilized the identity of the contested technology as a single artifact*, replacing it with multiple identities that have different levels of risk in specific situations or sites. One example is when actors enrolled geography, topography and climate as allies in order to distinguish between diesels in different areas (e.g. VV's, SNF's and Green Motorists' constructions of "city diesel" as different from "country diesel" – or Green Motorists' constructions of "southern diesel" conditions as different from "northern diesel" conditions). Numerous actors also constructed strong bifurcations between old and new diesel cars as having very different emission characteristics, as well as creating distinctions between "imported" and "domestic" diesel cars.

b   Actors *enrolled one's own (counter)expertise and enrolled global networks* as allies to strengthen their own networks (for example, BIL and VV asserted that "we are planning our own study"; Volvo Cars and BIL continually evoked the EU–ACEA agreement). This strategy also included efforts to *enroll technological progress* in general, as seen in the automobile industry's many references to "future technical improvements" in diesel motors as an important factor to be reckoned with.

The discursive strategies constituted an important means by which the various groups of actors sought to shape, contest, and control the content and outcomes of the risk discourse. As competing groups of actors put forth claims

and counterclaims about environmental risks, "facts" were reopened to negotiation, government policies were redefined, and transeuropean politics were
played out in global and local arenas. In the process, the artifact itself was
ascribed with shifting, heterogeneous identities and shifting politics.

Related to these dynamics, the third conclusion that can be drawn from
the controversy on the environmental risks of diesel cars is that this technology debate is indeed significant as a "site of political contestation" (Clarke
and Montini 1993). Claims about CO_2 emissions and assertions about technological advances, percentages and implications are intertwined with highly
charged, long-term political agendas in which the actors have high stakes.
Government agencies have high stakes in carrying out their long-term plans
and goals as mandated by the Swedish Parliament. Environmental groups
have strong commitments to ameliorating the environmental effects of automobility. Car manufacturers have high political and economic stakes to fulfill
the EU–ACEA goals and thereby avoid other, costly European regulation
that might threaten the long-term expansion of their market. It is clear that
the multiple constructions of the technology and its risks are inseparable
from its politics. Risks and politics are co-constructed in ongoing practice
and at multiple sites of contestation.

As vividly illustrated by the EU–ACEA agreement in which diesel cars
have a potentially important – but fiercely contested – role, the long-term
economic and environmental implications of ongoing debates about these
cars are significant. From a social science perspective, the debate shows the
various ways in which disparate groups of actors with diverse perspectives,
agendas and resources struggle to define, contest and stabilize risk constructions in their everyday practices. From a broader societal perspective, these
dynamics in turn vividly illustrate the strategies and politics by which such
actors negotiate and interact in ongoing struggles over the future of the car.

Notes

1 Although many of the actors who were interviewed in this study were typically
  experts within their respective organizations and influential in shaping their organizations' interpretations of the issues, they should not necessarily be viewed as official
  spokespersons for their respective groups. Obviously large organizations such as Volvo
  Cars and the Swedish Road Administration encompass many individuals with perhaps
  conflicting views on a given phenomenon. Interestingly, none of the people I interviewed made any comment to the effect that their views were not to be seen as
  synonymous with those of their respective organizations.
2 Lars Nilsson, director of Swedish National Road Administration, cited in Swedish
  National Road Administration (1999a: 7).
3 See Regeringsbeslut (Swedish government decision) M98/1689/7, miljödeparte-
  mentet (Ministry of Environment), 1998–05–14.
4 Specifically, SEPA's report estimates that annual CO_2 emissions from new diesel
  cars, based on 35 percent city driving and 65 percent country driving, are
  about 25.08 kg/car, while comparable emissions from new gasoline cars are about
  27.19 kg/car. See SEPA (1998: 6).

5 For example, in a 1997 article entitled "Diesel- or gasoline-fueled car?", the advantages and disadvantages of each type of car in terms of relative emissions of various substances are discussed. The advantages of a diesel motor are its low fuel consumption, its robustness for cold starts and various driving styles and its lower emissions of carbon dioxide. The disadvantages are that diesel cars are a major source of particle emissions: in a recent publication, VV claims that a diesel car produces approximately thirty times more particles than a gasoline motor equipped with a catalytic converter (see Swedish National Road Administration 1999a: 5).

6 Senior officer Olle Hådell, telephone interview, October 4, 1999.

7 Ibid., October 4, 1999.

8 Reino Abrahamsson, SEPA, telephone interview May 5, 1999 and personal communication February 3, 2000.

9 Ibid.

10 Lars Näsman, vice-chair BIL, telephone interview October 6, 1999.

11 Karin Kvist, environmental expert BIL, personal communication February 4, 2000.

12 A Volvo Group web page from 1999 claims that emissions of "legislated exhaust gases" from Volvo's diesel engines have dropped by 70–85 percent since the mid-1970s (Volvo Group 1999a). For more information on diesel, see Volvo Group (1999b).

13 Stephen Wallman, expert for mobility and technology assessment, telephone interview January 25, 2000; division chief Bo Ljungström, telephone interviews June 2, 1999 and October 4, 1999.

14 Specifically, Volvo's Position Statement notes:

> The prevailing methods used for particle measurements are based on total specific mass, and do not consider particle sizes. Analytical methods for particle size distribution in engine exhausts are not yet standarised, and several international development projects are going on. As the testing conditions affect the result, comparisons between different studies are difficult until the methods have been standardised.
>
> (Volvo Group 1999b: 1)

15 Division chief Bo Ljungström, telephone interview June 2, 1999.

16 Bo Ljungström, telephone interview June 2, 1999; Stephen Wallman, telephone interview January 25, 2000. See also Wallman (1998: 3).

17 Expert Stephen Wallman, Volvo Cars, telephone interview January 25, 2000 and personal communication February 3, 2000.

18 Telephone interview with Reino Abrahamsson, SEPA, September 28, 1999; telephone interview with Steven Wallman, Volvo Cars, January 25, 2000. Both participated in the hearing.

19 Reino Abrahamsson, telephone interview September 28, 1999.

20 Claes Pile, SNF traffic expert, telephone interview, June 1999.

21 Per Kågeson, Green Motorists, personal communication March 29, 2000.

22 I thank Per Kågeson, Green Motorists, for pointing this out to me.

23 Seminar "How dangerous is it to breathe?", organized by SNF, Stockholm, February 3, 2000.

24 Magnus Nilsson, head of SNF Stockholm. The claim was confirmed in a telephone interview, September 28, 1999.

25 This objective is cited in Tengström *et al.* (1995: 40). For a discussion of ACEA, see also Tengström pp. 40–41, 54–57.

26 Specifically, this level is 140 g/km (as compared with SEPA report's estimate of the current level of emissions in new cars of 183 g/km). The agreement also includes two interim goals, one of which is to reduce CO_2 emission levels to 165–170 g/km

by 2003 at the latest. See, for example, European Commission, notes from Council's 2,165th meeting – Environment – Brussels, March 11, 1999 ("Rådets 2165:e möte – Miljö – Bryssel den mars 11, 1999"), C/99/71.
27 Kurt Palmgren, CEO BIL, cited in BIL Sweden (1999b: 1).
28 Steven Wallman, Volvo Cars, cited in BIL Sweden (1999a: 4).
29 Steven Wallman, Volvo Cars, telephone interview, January 25, 2000.
30 For example, in May 1999 VV organized a "particle conference" in Stockholm; see documentation in Swedish National Road Administration 1999a. In February 2000 SNF organized a public seminar "How dangerous is it to breathe?", which attracted some seventy-five representatives from various organizations and the public.

References

BIL Sweden (1999a) *Bilbranschen just nu* ("Automobile Sector Right Now"), newsletter no. 1, January 1999.

—— (1999b) "CO$_2$-målet Kräver Diesel" ("CO$_2$ Goal Calls for Diesel"), in BIL Sweden (1999a) *Bilbranschen Just Nu* ("Automobile Sector Right Now"), newsletter no. 1, January 1999: 1.

—— (1999c) *Home Page.* Available HTTP: <http://www.bilbranschen.com/bilen/bil_trafik/text.asp?ID=8> (accessed June 1, 1999).

Bradbury, J.A. (1989) "The Policy Implications of Differing Concepts of Risk," *Science, Technology & Social Change* 14(4): 380–399.

Clarke, A. and Montini, T. (1993) "The Many Faces of RU486: Tales of Situated Knowledges and Technological Contestations," *Science, Technology & Human Values* 18(1): 42–78.

Dratwa, J. (2001) "Taking Risks with the Precautionary Principle," paper presented at workshop "Technologies of Uncertainty," Cornell University, April 20–22, 2001.

Green Motorists (1999) "Gröna Bilister utser 1999 års miljöbästa bilar" ("Green Motorists identifies the most environmental cars of 1999"), *Trafik & miljö,* 1: 9–15.

Hannigan, J.A. (1995) *Environmental Sociology: A Social Constructionist Perspective,* London: Routledge.

Hilgartner, S. (1992) "The Social Construction of Risk Objects: Or, How to Pry Open Networks of Risk," in J.F. Short, Jr and L. Clarke (eds) *Organizations, Uncertainties and Risk,* Boulder, CO: Westview Press.

Kågeson, P. (2000) *The Drive for Less Fuel: Will the Motor Industry be Able to Honour its Commitment to the European Union?* European Federation for Transport and Environment, T&E 00/1, January 2000.

Lenner, M. (1998) *Kvantitativ beräkningsmodell för PC för trafikens utsläpp av cancerframkallande ämnen i svenska tätorter* ("Quantitative Calculating Model for PC for Traffic Emissions of Carcinogeneous Materials in Densely Populated Swedish Areas"), Swedish Institute for Road Traffic Research, arbetsmaterial 1998; cited in Swedish Environmental Protection Agency, *Vad blir konsekvenserna för miljö och hälsa av en ökad andel dieselbilar i den svenska bilparken?* ("What will be the Consequences for Environment and Health of an Increased Percentage of Diesel Cars in the Swedish Car System?") dnr. 558–2542–98 Rr, October 12, 1998.

Lidskog, R., Sandstedt, E. and Sundqvist, G. (1997) *Samhälle, risk och miljö,* Lund: Studentlitteratur.

Nelkin, D. (1989) "Communicating Technological Risk: The Social Construction of Risk Perception," *Annual Review of Public Health* 10: 95–113.

Parthasarathy, S. (2001) "Knowledge is Power: Genetic Testing for Breast Cancer in the United States and Britain," paper presented at workshop "Technologies of Uncertainty," Cornell University, April 20–22, 2001.

Schmid, S. (2001) "Transformation Discourse: Nuclear Risk as Strategic Tool in Post-Soviet Politics of Expertise," paper presented at workshop "Technologies of Uncertainty," Cornell University, April 20–22, 2001.

Swedish Environmental Protection Agency, SEPA (1998) *Vad blir konsekvenserna för miljö och hälsa av en ökad andel dieselbilar i den svenska bilparken*, Redovisning av regeringsuppdrag (Results of Government Commissioned Study) M98/1689/7 (dnr. 558–2542–98 Rf), October 12, 1998.

Swedish government proposal (1997/98: 56) *Transportpolitik för en hållbar utveckling* ("Transport Policy for a Sustainable Development").

—— (1997/98: 145) *Svenska miljömål – miljöpolitik för ett hållbart Sverige* ("Swedish Environmental Goals – Environmental Policy for a Sustainable Sweden").

Swedish National Road Administration (1997) *Home Page*. Available HTTP: <http://www.vv.se/miljo/vagen_fram/allakan.htm> (accessed June 9, 1997).

Swedish National Road Administration (1999a) *Partiklar – trafik, hälsoeffekter och åtgärder: dokumentation från en konferens i Stockholm den 6 maj 1999* ("Particles – Traffic, Health Effects and Action Measures: Documentation from a Conference in Stockholm May 6, 1999"), written by Runo Ahnland for the Swedish National Road Administration, publication 1999: 98.

—— (1999b) *Bilars säkerhet och miljöpåverkan* ("Safety and Environmental Impacts of Cars").

—— (1999c) *Vilken bil väljer du? Kalla fakta om bränsleförbrukning, koldioxid och miljöklass* ("What Car do you Choose? Hard Facts on Fuel Consumption, Carbon Dioxide and Environmental Class").

Swedish Society for Nature Conservation (1995) *Home Page*. Available HTTP: <http://www.snf.se|verksamhet|trafik|dieselraps.htmp> (accessed 28 May 1995).

—— (1996) *Grön Väg: Idéer för hållbara trafiklösningar* ("Green Road: Ideas for Sustainable Traffic Solutions"), Stockholm: Naturskyddsföreningen.

Tengström, E., Gajewska, E. and Thynell, M. (1995) *Sustainable Mobility in Europe and the Role of the Automobile: A Critical Inquiry*, Stockholm: Kommunikationsforsknings-beredningen, p. 17.

Volvo Group (1999a) *Web Page*. Available HTTP: <http://www.bus.volvo.ses/corevalues/environment/engines.asp> (accessed May 11, 1999).

—— (1999b) Position Statement *Diesel: Engines, Fuels and Emissions of Particulate Matter*, reg. no. 99–164, 29 June 1999.

Wallman, S. (1998) Report *Kommentarer till Naturvårdsverkets rapport 98–10–05 – M98/1689/7* ("Comments to SEPA's report 98–10–05"), reg. no. 56020–109, 20 November 1998.

Wynne, B. (1989) "Sheep Farming after Chernobyl: A Case Study in Communicating Scientific Information," *Environment Magazine* 31(2): 10–15, 33–39.

Zwanenberg, P. van and Millstone, E. (2000) "Beyond Skeptical Relativism: Evaluating the Social Constructions of Expert Risk Assessments," *Science, Technology & Human Values* 25(3): 259–282.

5 Defining risk and safety in a high security organization

"Bunkering" at the Los Alamos Plutonium Handling Facility

Andrew Koehler

A variety of studies of the social construction of risk have discovered wide differences between the risk perceptions of those who work inside the dangerous systems of modern society and those outside (Bassett *et al.* 1996; Gregory *et al.* 1996; Jonas 1973; Kasperson *et al.* 1988; Kunreuther *et al.* 1994; La Porte 1996; La Porte and Metlay 1975; Mehta 1998; Plous 1989; Richards 1994). In *Risk and Culture*, Douglas and Wildavsky (1982) describe this difference as an almost Manichean clash of cultures between the "border" and the "center." Those who operate power plants, waste facilities, aircraft carriers or chemical plants often "have general confidence in the counteractive and anticipatory powers of their institutions and unlike the "boundary" risk perceptions of system outsiders may be disposed to estimate risk differently from those who mistrust their institutions" (1982: 89).

Observing this struggle between risk frames, Douglas and Wildavsky conclude, "If the selection of risk is a matter of social organization, the management of risk is an organizational problem. Since we do not know what risks we incur our responsibility is to create resilience in our institutions" (1982: 89). Others, such as Pool (1997), Hughes (1998) or Pinkus (1997) recommend similar basic proscriptions while emphasizing different aspects of "institutional resilience" such as public trust, designed flexibility, or improved technical decision making (Hughes 1998; Pinkus 1997; Pool 1997).

Yet despite the analysis of risk perception and system operation, relatively little has been written from the perspective of those inside various systems. Throughout the literature on risk perception, the focus has been either on how differences in risk perception affect individual engineer decision making or on explaining why "the public" opposes particular kinds of technical development. Less well studied are the longer-term effects that discrepancies in risk perception have on institutional characteristics and relations with the outside world.[1] On the scale of deploying a specific technical activity, incongruities in risk culture affect what design characteristics will be selected by designers and opposed by outside parties. At the level of the institution, incongruities in risk culture also play a role in determining how the organization as a whole sees itself, how it relates to the outside world, and how it carries out long-term planning. From the perspective of public policy, understanding these "incongruity effects" is important: a technical organization that sees itself as doing work that is not widely understood or appreciated may develop characteristics over time that may erode its ability to function safely or effectively.

In this chapter, I will use the example of the US Plutonium Handling Facility at Los Alamos National Laboratory, New Mexico, to explore how risk is managed in a facility facing a divide between internal and external perceptions of risk. The facility, known as "TA-55," is described as the "nation's sole remaining full service plutonium laboratory" and employs approximately 750 people. I conducted qualitative fieldwork at Los Alamos in 1998 and 1999, and this chapter is based on my interpretations of a variety of conversations workers were willing to have about their lives and the demands placed upon them in the service of the US (see also Koehler 2001).

Used both for research into actinide chemistry and weapons stewardship, TA-55 routinely performs tasks that, if not done properly, could result in the death of workers, exposure of the off-site public to radiation or hazardous chemicals, or the destruction of expensive laboratory infrastructure. To control these risks, and to satisfy statutory requirements, approximately 100 full-time employees are employed either to ensure that the facility is operating safely or to ensure that it will perform as desired in the case of an accident. Much of the facility's $175 million dollar annual operating budget is devoted to either safety or facility improvements (Kim 2000). All facility workers are trained to follow a wide gamut of rules and advisories varying from "Site Specific Training" to "Work Smart Standards" to "Equipment Qualifications." Plutonium workers have additional safety training requirements that can absorb up to several months per year. Yet despite the considerable resources devoted to the idea of safety, choices still must be made between which risks will and will not be considered. Wildavsky and Douglas could be speaking about TA-55 when they write, "Where does the path of virtue and good sense lie . . . since it is not possible to say everything at once, or with equal emphasis, some sort of risk has to be hidden" (1982: 27).

From the perspective of the literature on social construction of risk, many of the factors influencing how, and which, risks are identified and mitigated are plainly apparent at TA-55. Much of what determines the importance of a particular risk does indeed depend upon cultural and social factors such as the degree to which workers share a common belief in the facility mission or the relative differences in power between subgroups within the organization that operates TA-55. However, in the case of TA-55 boundaries play an especially large role in creating the risk agenda "inside" the facility. Boundaries also play a key role in creating specific operational behaviors, one of the most interesting of which I have labeled "bunkering."

Risk perception and boundaries

That participants in a technical system tend to develop a different understanding of system risks than constructed by those outside system boundaries is a well-established topic (La Porte 1996; Perrow 1984; Slovic *et al.* 1994). In the case of TA-55 these differences are especially dramatic. Even among nuclear facilities, TA-55 is an extreme case: in the Department of Energy's classification scheme, it is a "Category-I" facility amply equipped with gates, guards and guns (known as "triple-g"). The area surrounding TA-55 requires a security clearance to enter, and a sizable security force is equipped with machine guns and armored vehicles to protect the several metric tons of weapons-grade plutonium kept on-site. Razor-wired topped fencing marks both the physical and mental boundaries of the facility. At night, high-powered sodium lights bright enough to be seen in the city of Santa Fe (35 miles away) flood the facility with yellow beams. Every few months local papers report that some program with an acronym like ARIES or MOX will be worked on at the facility. Otherwise, the most visible activity of the facility seen from outside consists of tinted-windowed suburban vans driving about the perimeter fence.

From the perspective of the average outside viewer, TA-55 is the very vision of a suspicious government facility. To those already opposed to US nuclear policy, TA-55 is mentally akin to Dracula's castle on a mesa – with the exception that TA-55 has been described as "containing enough deadly poison to kill everyone on Earth." Other observers are reminded of images from films like *Dr No* or *Men in Black*; indeed a rumor in Santa Fe has it that alien autopsies are part of the Lab's efforts.

For those attached to the facility, TA-55 appears differently. Through orientation training, badge processes and technical briefings, new employees are quickly acclimated to the unusual aspects of the facility. On a day-to-day basis, most of what goes on inside is mundane. Like any office, paperwork has to be filled out, human resource reports filed, computer hassles fixed and office supplies located. Although plutonium handling requires special care, many of the general activities at the facility are similar to those conducted at any light machine manufacturing plant: welders, electricians and tin-workers install and remove equipment, leaving at 5 o'clock with lunch pails in hand.

Plutonium work is not a part of most workers' tasks – and for those employees for whom it is, plutonium represents one set of hazards among many. Certainly, great emphasis is placed on mitigating exposure to plutonium; however, on a daily basis workers also encounter hot furnaces, electrical equipment, acids, and other hazards. In terms of the direct hazards to worker health, plutonium is considered to pose a set of unusual but controllable risks.

No one "inside" the facility would disagree with the statement that plutonium is inherently dangerous. But while an "outsider" might interpret the imposing physical controls that contribute to the visible appearance of TA-55 as evidence that plutonium handling is something worse than snake handling, to an "insider" these controls are symbols of service being performed in the service of national policy. An ability to maintain security, to quietly go about business and to maintain the environmental and security "walls" keeping plutonium apart from the public is not a symptom of nefarious purposes but rather proof of honorable intention.

To the "insider," there should not be a need for public concern about plutonium operations when TA-55 is operating as designed. The considerable expense, effort and disruptions to work made by radiation, material accountability, security and environmental controls are external signs that TA-55 is capable of controlling the risks of plutonium. If the security precautions that make up part of these systems are imposing and ominous, then it should be obvious that they are effectively deterring theft of "special nuclear material" by terrorists, rogue nations or the deranged.

Bunkering

The differences that such striking facility boundaries promote between these two different risk perspectives are interesting but not operationally significant just because they exist. They become significant, however, through the tendency of such differences in risk perception to promote a kind of "bunker mentality" or "bunkering" among those involved in design and risk mitigation tasks "inside" the facility.

Because of the fundamentally different perceptions about the risks posed by the facility and the nature of the boundaries between TA-55 and the outside world, communication is guarded. Bound by security restrictions and the fundamental divisions between how the facility looks from the "inside" and "outside" perspectives, those working at TA-55 have difficulty explaining what they do to the wider world. In part this difficulty is caused by workers' fears that they will meet with a hostile response; in part it is due to institutional restrictions. However, a significant divisive mechanism at work is cultural difference.

When someone spends so much time in a relatively isolated social environment – one with its own terms, modes of expression and common understandings – it often requires considerable effort to express ideas to those outside this environment. This situation is especially the case with regard to risk communication. For example, in order to explain from the perspective of TA-55 why public fears over an upcoming fuel shipment are misplaced, a host of "insider" concepts have to be explained such as:

- how Secure Safe Transport shipments operate;
- the nature of 3013 shipping containers;
- the amount of material at risk and why a criticality-safe container works;
- the track record of similar shipments over the past four decades;

- relevant environmental, health and safety methods and standards;
- the atomic and chemical nature of the materials and the methods of stabilization.

If these concepts and procedures are not familiar to the academic reader, they are even less familiar to the general public. The average TA-55 worker does not have time or the energy to explain the "inside" culture to academics, much less to the general public, given that the facility's public relations specialists have generally failed to develop effective external communications. There are many other less taxing things to do at the end of the workday than explain (or possibly defend) facility activities to family, neighbors or the curious. Keeping track of exactly what information connected with facility operations may be sensitive is another drain on mental energy: it is simply far easier not to say anything at all than risk being disciplined for a security infraction.

Because the cultural divide over risk and other aspects is so great between the inside and the outside, members of the facility withdraw into their own group. These cognitive and physical separations help foster a culture within TA-55 that can resemble an insular ethnic group's cooperative organization (for example, the mutual aid societies in San Francisco's Chinatown). Family members tend to take on the same trades, and workers tend to socialize "in group." Indeed, the rates at which more than one family member works at TA-55 seem (based on anecdotal evidence) high compared to other parts of either Los Alamos or the public sector. Office marriages and romances also occur, and the sight of a couple giving each other a peck on the cheek before both go to work at the Lab is not unusual – except that in one case I observed, both partners were camouflage-clad and carrying automatic rifles.

The insular culture and the withdrawal from the outside – which I am calling "bunkering" – is, however, not a paranoid or hostile response to feeling misunderstood or somehow different from the rest of the world. Workers often understand why "outsiders" in Santa Fe do not look at risks the same way as they do or even why some "outsiders" vehemently oppose TA-55's activities. Given administrative controls on what can be discussed, and the mental costs of trying to explain differences in risk perception, however, it is far more enticing to remain out of the line of fire and with a group of people who already understand the facility's risk framework. Unfortunately, by "bunkering," staff and managers remove themselves and their potential input from the political and social decision-making processes that determine TA-55's fate.

Bunkering and its impact upon risk management

While bunkering may be a natural response to the environment and a means of conserving cognitive energies, it has implications for how facility managers handle risk and operational activities.

Risk analysis and the role of tradition

At a facility like TA-55, risk is managed through procedure. When plans call for a new process (for example, a glove-box line for fuel fabrication) the Hazard Analysis (HA) group at the facility is notified by the design team of the proposed modifications. The HA group compares the proposal against two criteria: first, how does the proposed change affect the overall risk profile of the facility, and second, how does the proposed change affect worker health and safety? Among these two criteria, the first is the most important in statutory terms. By the terms of its operating license, TA-55 must maintain the risk of an accident with off-site consequences at a fixed probability. If changes are deemed not to have a significant impact on the overall level of facility risk, they are approved and released for work. Changes that increase the facility risk envelope or involve alterations to safety class equipment are, however, categorized as "Unresolved Safety Questions," which in turn requires that a formal probabilistic risk assessment (PRA) be performed before they can be "resolved."

PRAs are both difficult and expensive. Furthermore, in order to calculate PRA accident probabilities, scenarios must be developed to describe the various possible failures of equipment in the facility. Since it cannot be proved that all possible means of failure have been explored, tradition plays a strong role in determining what accident scenarios will be considered. Engineers calculating the "risk contribution" of a new process generally rely on analyses that TA-55 has used in the past to evaluate analogue processes. These scenarios have been codified through a series of checklists that provide guidance as to what factors should be considered. For example, if a process requires a tank of fluid, PRA checklists will recommend that the possibility of spillage due to seismic activity should be analyzed.

Many of these "standard" accident scenarios are specified by the facility's license or some other statutory requirement. Some scenarios exist, however, because at some point in time it was administratively decided that they should. Others were developed from the set of accident scenarios that were developed for regulation of the nuclear power industry. For these latter two categories of accident scenarios, HA staff cannot always know the requirements pedigree behind the scenario, and the scenarios themselves may or may not apply well to a non-reactor research facility such as TA-55. Nevertheless, there is a bias towards reusing accident scenarios that are already a part of TA-55's standard set rather than to develop new scenarios, because (a) the traditional scenarios have already been found to be acceptable by regulators, and (b) the analytic solution methods (generally an analytic or simulation model) are understood by TA-55's management.

Bunkering contributes to the strength of this tradition in scenario choice primarily because it reinforces the sense that the facility must rely on its own solutions and means of evaluating risk. I am not qualified to assess whether or not the accident scenarios presently in use are the right ones, and TA-55 has an excellent safety record compared to other Department of

Energy facilities. However, when charged with studying a new process typically, in the words of one HA specialist, "we will use what we know or guidance from somewhere else in the nuclear world." As most activities at the facility do not involve nuclear materials, there is a wealth of other kinds of risk assessment methods from the chemical and biochemical industries that might prove useful. Because these methods are from "outside" sources, however, from the vantage point of TA-55 they have not been pursued – at least in part due to fear that changes might promote doubt among external parties over the legitimacy of the existing HA traditions.

Bunkering and the perception of politics and risk analysis

The politics of risk regulation are undeniably complex at a facility such as TA-55. At present, TA-55 is overseen by the US Department of Energy, although the Nuclear Regulatory Commission may take over the primary regulatory role at some time in the future. In addition to the Department of Energy, TA-55 must also cooperate with regulators from the State of New Mexico, other federal agencies such as the Environmental Protection Agency or the Department of Transportation, local municipal bodies and the surrounding tribal pueblos.

With so many different parties playing a role in defining the acceptable operating envelope, facility managers understand all too well that the risk standards enforced by regulators result from political processes. For example, when the Department of Energy first proposed sending shipments of low-level nuclear waste to a site outside Carlsbad, New Mexico, state officials faced public pressure to regulate radioactive shipments more tightly. In turn, New Mexico officials responded by demanding that the Department of Energy provide new roads so that trucks would be routed around Santa Fe. Likewise, standards over radiation doses to the public typically are reconsidered every time there is a prominent nuclear incident such as those at Chernobyl or Three Mile Island (Rees 1994). Yet at the same time that TA-55 is forced to operate under *politically established* risk standards, managers at the facility face pressures to act as though the risk analyses resulting from those risk standards are *scientific and impartial*.

Internally, the pressure for scientific PRA results stems from programmatic factors. Program schedules depend, for example, upon completing Environmental Impact Statements. Similarly, jobs, budgets and the support of those higher in the chain-of-command depend upon whether a particular activity contributes one in 10^9 latent cancers per year or only one in 10^8 such cancers. On a daily basis, the mode at which the facility is allowed to operate is determined by how TA-55's risk profile compares to key risk/hazard indexes. Should the facility's risk profile rise too high, managers may be forced to stop activities, or even close the entire facility. From a management perspective, therefore, the exact calculated value of process risk at the facility matters a great deal.

Externally, pressures for scientific PRA findings are exerted from the chain of command. The Department of Energy is responsible for ensuring that programs planned for TA-55 are compatible with legal requirements for public safety. As a program works its way through the budgeting process, risk and hazard values play a key role in the battles that ensue. Opponents of the program can often delay or derail a program entirely if they are able to show that calculations of public safety are flawed. Program supporters, on the other hand, must have a consistent, supportable and defensible set of values by which they can convincingly argue that the safety of the public is being assured. These values must not be seen to change over the course of the program – indeed the worst possible outcome is for supporters to be forced to revise their own safety values (as has occurred several times in the ill-fated Yucca Mountain high-level waste repository project (Arrandale 1997; Easterling and Kunreuther 1995)).

Caught between on the one hand, an inherently political process driving the need to analyze particular risks, and on the other hand, a second set of forces which demand that the results of risk analyses be scientific (yet support program objectives) risk managers thus face a conundrum. The problem is that risk analysis, particularly that which involves low probability/high hazard activities, is inherently imprecise, and there are risk analysis methods that give results that are clearly wrong. There are also risk analysis methods that are "accepted" in that they are rigorous and consistent with physical laws. The kinds of risk analysis that are used most often at TA-55, however, cannot be said to be scientific simply because the results from these analyses are not reproducible.

This problem of reproducibility in PRA analysis is fundamentally insurmountable in the real world: there are no metrics that can distinguish which "acceptable" risk analysis method yields the true result from the set of all "acceptable" risk analysis methods. Since many of the Hazard Analysis scenarios are of extremely low probability they may rarely (and perhaps never) have occurred under any circumstances, much less under the exact circumstances that exist at TA-55. Frequently, therefore, events must be modeled for which little or no data exist in the real world that can be used to validate results. In the absence of actuarial data or metrics to judge between "acceptable" models, some kind of rule – either arbitrary, random or based on preferences – must be used to select either a methodology or a result.

Furthermore, even if actuarial data on seismic events at plutonium facilities existed, a host of other factors introduce irreproducibility into efforts at risk analysis. Typically, risk calculations must be performed for programs before either the equipment being deployed or the exact procedures to be followed are determined. In cases where risk managers can obtain master equipment lists and operating procedures for such projects, results may vary due to disagreement over what happens during a particular accident scenario or because different physical assumptions of how various facility processes work are used. To make matters worse, even when identical physical assumptions are made, different modeling codes will often yield answers that vary by powers of ten. Because such programs are themselves built from complex webs of assumptions, determining whether these discrepancies are due to methods of handling small probabilities, rounding errors or method-ological differences is a challenge. Finally, risk managers understand all too well from experience gained from other PRA studies that there is no way to ensure that the scenarios *omitted* from analysis are not more impor-tant than the those that are *included* (US Nuclear Regulatory Commission 1975).

In the face of this tension between needing exact results and having to use inexact techniques, the "bunkering" response plays a role in encouraging a withdrawal from either playing a role in setting the external risk agenda or in communicating with the public more openly about program risks. Although managers at TA-55 know that risk analysis is more art than science, the vast cultural divide between "inside" and "outside" – combined with the institutional pressures – work to encourage the presentation of risk analyses as being impartial and precise. In the past, TA-55's management has consid-ered increasing their communication with the public about operations and safety, but there are deeply ambivalent feelings about such efforts. As one manager put it, "if we stick our necks out, talk safety and do [public] outreach, we'll just get hammered."

Bunkering and design choice

Bunkering also plays a role in determining the technical choices that are made by designers of both safety and process equipment. Two examples of the impact of bunkering on design are given by the deployment of light-ning protection and a recent program to fabricate fuel pellets.

Lightning protection

One of the most striking exterior features of the TA-55 facility is an impres-sively large number of lightning rods covering the roof – about one lightning rod every three square meters. Although lightning strikes are a frequent and dramatic occurrence in New Mexico, no other building I have ever seen has as many lightning rods as does TA-55. These rods are a relatively new addition and, notably, were not installed because calculations indicated a particular vulnerability to lightning. Rather the rods were the result of an inspection by a congressionally appointed oversight group that periodically inspects TA-55. Members of this group come from other nuclear facilities or academia, and they typically are chosen for their reputation as experts in particular technical fields.

The year that the lightning rods were deployed, someone with a partic-ular interest in electrical safety led the inspection team. Not surprisingly, the team's recommendations included advice on strengthening TA-55's lightning defenses. Given the choice between disputing the recommendations of a publicly appointed oversight group or deploying what was felt to be an overabundance of equipment, TA-55's managers chose both the rational (in my view) and culturally expectable solution.

Fuel fabrication

Similarly, since 1979 the US has maintained a policy against "closing the fuel cycle." This means that the US has not developed the capability to reprocess civilian nuclear fuel: once reactor fuel completes a scheduled burn-up, the spent fuel is disposed of rather than used to extract the remaining useful uranium and plutonium from the spent fuel so that it may be reused. The standard method of reprocessing that is used around the world is based on a process called PUREX that was developed in the US in the 1960s. At the heart of this process, fuel rods are chopped up and the fuel is dissolved in acid. Later, through the use of a variety of chemical techniques, the useful uranium and plutonium in the acid solution are extracted and purified. Motivating the decision to only burn fuel "once-through" was the fear that the spread of reprocessing technology would lead to greater proliferation of nuclear weapons. States with the ability to perform reprocessing have the ability to make weapons-grade plutonium. The US hoped that by setting the example of not reprocessing civilian nuclear fuel, other nations would also choose not to do so – thus preventing the spread of weapons technology (Rochlin 1979).

The legacy of this policy affects TA-55 in an unexpected way. As part of its research into nuclear reactor fuel design, TA-55 fabricates fuel pellets containing plutonium. Fuel fabrication is tricky, and the pellets must meet very strict physical requirements. One such requirement is that the pellets must be sintered properly so that, when irradiated, they will not crumble or develop cracks inside the fuel cladding. In order to determine whether or not a batch of pellets meets this requirement, the simplest testing method involves placing a pellet in an acid solution to observe how it dissolves.

By some interpretations, this method of acid testing might be a symbolic step towards resurrection of the reprocessing "ghost." The fear that external groups might misrepresent TA-55's use of this test as an attempt to run afoul of US reprocessing policy recently became an issue to a team of designers on a project in which the sintering quality of the pellets needed verification. Ultimately, considerable time and effort was devoted to studies of alternative techniques and the possibility of relaxing the requirements. The sense among the design team was that while they did not know if anyone would raise objections to the acid testing method, by performing alternative studies the decision whether or not to use the acid technique would be "bumped to a higher pay-grade," that is, delegated to higher authority. In this case, bunkering contributed to a programmatic delay by increasing the uncertainties with regard to equipment choice, schedule and budget.

Although TA-55's program ultimately made use of the acid testing technique (which is also routinely employed by US commercial and foreign fuel fabrication organizations), the degree of expressed concern about the possibility of external action was remarkable. A decision that in France, Britain, or Belgium would be considered a routine technical one became a programmatic obstruction in the US based on the probably slim possibility of having to face a specific opposition strategy.

Conclusion

A plutonium handling facility such as TA-55 must maintain rigid boundaries both mentally and physically between itself and the rest of the world. In the US nuclear political environment, one of the costs of maintaining these boundaries is a kind of withdrawal from the outside world on the part of those "inside" the facility. Faced with the difficulty in the post-cold war era of describing the facility's mission, as well as differences in risk culture, limits on speech and the perception of exterior indifference or hostility, workers "inside" the facility have incentives to avoid communication. They essentially "bunker down" and hope that the "outside" does not inflict anything too terrible.

While such a (largely unconscious) strategy is understandable, it results in costs that are incurred both inside and outside the plutonium facility. Because those inside the system do not develop greater ties to the public, to regulators, and to policy makers, they are not able to participate in political processes as effectively as they might need to in order to protect the interests of the facility. These dynamics result in a situation in which the facility is continually reacting to external forces – either real or feared – rather than being able to plan for the long term. For external actors as well, TA-55's "bunkering" incurs costs in the form of increased fear and mistrust in TA-55 and government in general, in not being provided with as much information as possible, and in a reduced likelihood that the facility's choices with regard to risk and operations will be satisfactory.

Note

1   A notable exception is Gusterson (1998).

References

Arrandale, T. (1997) "The Radioactive Hot Potato," *Governing Magazine* November 1997: 68.

Bassett, G.W., Jenkins-Smith, H. and Silva, C. (1996) "Attitudes and Perceptions Regarding On-site Storage of High Level Nuclear Waste," *Risk Analysis* 16(3): 309–320.

Douglas, M. and Wildavsky, A. (1982) *Risk and Culture*, Berkeley, CA: University of California Press.

Easterling, D. and Kunreuther, H. (1995) *The Dilemma of Siting a High Level Nuclear Waste Repository*, Boston, MA: Kluwer Academic Publishers.

Gregory, R., Brown, T.C. and Knetsch, J. (1996) "Valuing Risks to the Environment," *Annals of the American Association of Political and Social Sciences* 545: 54–63.

Gusterson, H. (1998) *Nuclear Rites: A Weapons Laboratory at the End of the Cold War*, Berkeley, CA: University of California Press.

Hughes, T.P. (1998) *Rescuing Prometheus*, New York, NY: Pantheon Books.

Jonas, H. (1973) "Technology and Responsibility: Reflections on the New Tasks of Ethics," *Social Research* 40(1): 31–54.

Kasperson, R.E., Renn, O. *et al.* (1988) "The Social Amplification of Risk: A Conceptual Framework," *Risk Analysis* 8(2): 177–187.

Kim, K.C. (2000) *Nuclear Materials Technology Division Organizational Self-Assessment*, Los Alamos, NM: Nuclear Materials Technology Division, Los Alamos National Laboratory.

Koehler, A. (2001) *Design for a Hostile Environment: Technical Policymaking and System Creation*, Ph.D. dissertation, Goldman School of Public Policy, Berkeley, CA: University of California.

Kunreuther, H., Slovic, P. and MacGregor, D. (1994) *Risk Perception and Trust: Challenges for Facility Siting and Risk Management*, Philadelphia, PA: Wharton Center for Risk Management and Decision Processes.

La Porte, T. (1996) "Hazards and Institutional Trustworthiness: Facing a Deficit of Trust," *Public Administration Review* 56(4): 341–347.

La Porte, T. and Metlay, D. (1975) "Technology Observed: Attitudes of a Wary Public," *Science* 188(11 April): 121–127.

Mehta, M.D. (1998) "Risk and Decision-making: A Theoretical Approach to Public Participation in Techno-scientific Conflict Situations," *Technology in Society* 20: 87–98.

Perrow, C. (1984) *Normal Accidents*, New York, NY: Basic Books, Inc.

Pinkus, R.L., Shuman, L.J., Hummon, N.P. and Wolfe, H. (1997) *Engineering Ethics: Balancing Cost, Schedule, and Risk – Lessons Learned from the Space Shuttle*, Cambridge, MA: Cambridge University Press.

Plous, S. (1989) "Thinking the Unthinkable: The Effects of Anchoring on Likelihood Estimates of Nuclear War," *Journal of Applied Social Psychology* 19(1): 67–91.

Pool, R. (1997) *Beyond Engineering: How Society Shapes Technology*, Oxford: Oxford University Press.

Rees, J.V. (1994) *Hostages of Each Other: The Transformation of Nuclear Safety Since Three Mile Island*, Chicago, IL: University of Chicago Press.

Richards, M.D. (1994) *Making Up Our Mind in a Democratic Age: A Review of the Social Science Literature on Land Use Decision-Making*, Quebec City, Quebec: Nuclear Energy Institute.

Rochlin, G. (1979) *Plutonium, Power and Politics*, Berkeley, CA: University of California Press.

Slovic, P., Flynn, J. and Gregory, R. (1994) "Stigma Happens: Social Problems in the Siting of Nuclear Waste Facilities," *Risk Analysis* 14(5): 773–777.

United States Nuclear Regulatory Commission (1975) *Reactor Safety Study: An Assessment of the Accident Risks in U.S. Commercial Nuclear Power Plants*, WASH-1400. Washington, DC.

Part III

Constructing safety

Introductory comments

The chapters in Part III explore the processes by which risks are handled and safety and trust are mutually constituted in everyday sites of technological practice such as air traffic control centers, intensive care units, seaborne aircraft carriers, engineering firms and other workplaces. How are "cultures of safety" achieved, and how is trust developed and sustained in interactions among practitioners in high-risk, high-tech environments? How do power relations influence the ways in which uncertainties are handled among various groups? What characterizes the learning processes by which safety continually evolves in specific workplaces? The authors in this section have carried out extensive ethnographic work at the various sites, using this work as the basis for conceptualizing new understandings of risk and safety.

Gene I. Rochlin explores the interactive dynamics by which operational safety – or more accurately, "safeness" – is achieved in complex, high hazard organizations and systems, focusing on the importance of collective agency in providing such safety. Drawing upon the in-depth work of the Berkeley group on High Reliability Organizations (HRO, see the introduction to this volume), Rochlin argues that a collective commitment to safety is an institutionalized social construct – and "a story that a group or organization tells about its self and its relation to its task environment." Much more than a set of observable procedures or scripts, the means by which organizations sustain high levels of operational safety embody rituals, interactions and myths. Organizations that perform well weave stories and rituals into their operations; these stories and rituals serve to orally transmit operational behaviors, group culture and collective responsibility. The constructed narrative or "culture of safety" that results is a dynamic, intersubjectively constructed belief among operators in the potential for continued safety in carrying out their operations.

In order to maintain such safety, however, operators and managers must constantly engage in processes of learning. The issue, Rochlin argues, is not how actors come to know what they know, but rather how "the known" is interactively constructed in the process of knowing. This process involved both *learning as practice* and *practice as learning*. Rochlin analyzes a number of means by which learning takes place and safety emerges in interactions among actors who are situated in multiple sites within organizations. Among else, Rochlin observes that one specific mechanism is the "talk" among multiple communication links that operators typically engage in, both during critical operations and in the less intense – but far more common – everyday situations of low activity. A constant stream of talk is a means of maintaining the integration of the collectivity and reassuring all participants of its status. Safety becomes an emergent property that evolves at multiple sites and through multiple forms of agency.

The role of "talk" in constructing and maintaining safety in high hazard, sociotechnical systems is also an important theme in Johan M. Sanne's chapter on how trust is created and sustained in everyday interactions among air traffic controllers. Sanne argues that the risks of air traffic control systems are to a great extent handled through the ongoing production and acknowledgment of trust among air traffic controllers, as well as between controllers and pilots. Two key resources for creating trust are *common orientation* to the task at hand and *redundancy* in the form of overlapping knowledge and awareness of each other's tasks. By common orientation is meant both a common understanding of the situation at hand and a commitment to the team and the task: operators engage in collective sense making about the system as a whole and their role in it. Similarly, redundancy in the form of overlapping knowledge of each other's tasks is crucial to establishing mutual support. Sanne argues that common orientation needs to be continually demonstrated and acknowledged in actors' everyday communicative practices, where "radio talk" between air traffic controllers and pilots plays an important role.

Sanne goes on to show that within communities of air traffic controllers, there are at least three concrete mechanisms by which trust is created: through negotiations about the handling of specific aircraft, through coordinating and "reading" each others' actions in everyday work and through using technical means (such as images on radar screens) to monitor and support each other's work. The achievement of safety thus embodies ongoing *negotiations of trust* in specific forms of collaborative and communicative practices.

Sabrina Thelander discusses the processes of creating safety through negotiations that are considerably more problematic, namely negotiations among actors who are situated in different positions within strong hierarchies of power. Rather than reflecting means of creating trust, the dynamics in this case often concern how *distrust* toward a co-worker and her/his abilities must be handled – and in some instances hidden – as a means of maintaining safety. Thelander has studied an organization that is characterized by considerable unpredictability, complexity and uncertainty, namely a cardiac intensive care unit in a high-tech university hospital. The sources of this complexity and uncertainty are multiple: there is always a risk that the patient's condition will change rapidly (and might lead to death), staff must manage many types of sophisticated medical artifacts (which place demands on expertise), many different people with different areas of responsibility are involved in providing care, and quick judgments and decisions are often required in critical situations. For the individuals and groups involved, handling these circumstances requires not only medical knowledge and expertise but also social skills and the ability to use a range of technological devices in often stressful (and sometimes fearful) interactions with both co-workers and the bodies of patients. Safety is established by acting so that the unit is represented as a safe place and the individuals who work there as competent.

Inspired by the work of Goffman (1959) and Strauss *et al.* (1985), Thelander analyzes how safety is constructed in narratives and actions among licensed practical nurses, registered nurses and physicians as they carry out the "trajectory work" of providing care to each patient. Specifically, Thelander shows how nursing staff manage risk and certainty in two specific situations, namely when someone shows fear of making a mistake and when someone distrusts a co-worker's abilities to handle the technology or the situation at hand. An important finding of her analysis is that existing power hierarchies are reproduced in the practices by which uncertainties are handled and safety is created.

Finally, the chapter by Marianne Döös and Tomas Backström focuses on how "safety work" is carried out at various sites of modern working life in Sweden, particularly firms and factories. In recent years, increasing demands for *control*, on the one hand, and *learning and acting competently*, on the other hand, have appeared to come into conflict for organizations, their managers and their workers. One consequence of these changes is that safety at the workplace is increasingly constructed in the day-to-day performance of work tasks in ways that cannot be "managed" by targeted safety work alone. Two potentially incompatible approaches to maintaining safety have emerged – namely learning versus control as the basis for safety work. Workers are often caught between conflicting demands for the quite different abilities, actions and ways of thinking that these approaches imply. Despite these cultural incompatibilities, the point of departure for Döös and Backström's work is that handling risks through both control and learning is necessary in order to achieve and maintain a safe work environment. Like Rochlin, they ask what characterizes safety culture and safety learning: What are the mechanisms and situated interactions by which safety is collectively and reflexively learned within organizations? And similarly to Sanne, Döös and Backström observe that achieving safety culture is a process of collective learning in which shared knowledge and understandings, as well as collective accommodations, are important dimensions in maintaining safety.

Döös and Backström explore the potential in actual practice for coexistence of *risk control* and *safety learning* as strategies for creating safety in various organizational settings, drawing upon empirical data from two companies, a newspaper printing plant and a large engineering firm. These cases and other research (Döös 1997) show that high levels of workplace safety do not emerge from specific safety measures, but rather from the way in which people solve ordinary work tasks and their related problems. The case of the engineering firm, in particular, indicates that successful accident prevention can be achieved when participation and learning are added to a traditional safety culture that is based on measures of internal control. The question of whether an approach based on the opposite type of dynamics — that is, situations in which a control approach to safety is adopted within an approach based on learning — would work is still, however, an unresolved issue.

References

Döös, M. (1997) *The Qualifying Experience: Learning from Disturbances in Relation to Automated Production*, Dissertation, Department of Education, Stockholm University. (In Swedish with English summary.) Arbete och Hälsa 1997: 10, Solna: National Institute for Working Life.

Goffman, E. (1959) *The Presentation of Self in Everyday Life*, Harmondsworth: Penguin.

Strauss, A. *et al.* (1985) *The Social Organization of Medical Work*, Chicago: The University of Chicago Press.

6 Safety as a social construct

The problem(atique) of agency

Gene I. Rochlin

Studies of "risk" have become something of a cottage industry both in the US and in Europe in recent years. Instead, a better model might be something more Gesellschaft and less Gemeinschaft than the homely cottage industry, and of a more appropriate scale – General Motors perhaps, or Saab. As Ulrich Beck and other observers have so forcefully pointed out, modern societies have become almost obsessed with the question of risk and how to reduce or control it (Beck 1992; Beck 1996; Beck *et al.* 1994; Luhmann 1993). This emphasis has in turn spawned numerous modes and methods for analyzing and understanding risk, ranging from highly sophisticated computer models that measure the probabilities and consequences of equipment failures in the most exquisitely complex of technical systems (e.g. Freudenberg 1988; Ralph 1988) to studies of the cultural basis for acceptance of such familiar risks as hunting, childbirth and warfare (Douglas and Wildavsky 1982; Rayner 1992). And each of these domains is more and more heavily populated from year to year as our obsession with risk – and its assessment, reduction, mitigation and management – continues to grow (National Research Council 1996).

In looking over this academic bricolage in search of some systematic way to teach it, I noticed two things that stood out prominently in my mind. First, what are generally known in the trade as "risk studies" remain largely event- and process-driven. The *construction* of risk and safety tends to be the provenance of social theorists rather than risk analysts. For the most part, even those studies that profess to examine the "social construction" of risk tend to argue about how the "public," or the affected population, socially construct their perception of the probable harm from a specific activity or set of activities (Freudenberg 1992; Kasperson 1992; Renn 1992). There is an almost complete disconnect between, on the one hand, the mainstream risk literature, which parses risk by categories, sometimes differentiated by technology and sometimes by social or political practice, and on the other hand constructivist literature, or even cultural studies, let alone the highly integrated notion of risk and culture that underlies social theories such as Ulrich Beck's conception of a "risk society."

The second thing that stood out in my perusal of risk studies was the seeming acceptance of the notion that there was no real definition, observation, or social indicator called "safety" that was independent of thinking about specific risks or sets of risks (Rochlin 1999). It reminded me very much of the "modernist" literature on medical care, which defined health operationally in terms of being free from disease or other negative interventions. Only in recent years has the definition of a "healthy" individual expanded beyond what is essentially a list of mechanical indicators such as low cholesterol, clear skin, good vision and disease negations such as no sign of tuberculosis, hepatitis, infections, etc. to a positive construct that involves not just how a person "is," in the objectivist, positivist or modernist sense, but also how the individual "feels," that is, how he or she perceives his or her own state of physical and mental well-being. Many public health officials have recognized that deconstructing the generally held notion of health is far too difficult; instead, many have come up with the grammatically offensive but descriptively apt term "wellness" as an alternative (Edlin *et al.* 1996).

Perhaps we might follow the cultural theorists, who make a distinction between risk and danger (Rayner 1992), and argue that a similar distinction needs to be made between safety and something called "safeness." It is a horrible term, of course, epistemologically challenged and ontologically primitive, but it does avoid the fault of treating safety (like health) as the property of an isolated, unsocialized individual. In that sense it is closer to the anthropological use of the term "risk set" in Douglas and Wildavsky's *Risk and Culture* than it is to either the more instrumental definitions used in the prevailing literature or the social theoretic formations used by Beck *et al.* In a word (perhaps some *other* word?), it closely resembles middle-range theorizing on a social scale and is therefore most suggestive of how a typology might be constructed that incorporates the notion of agency at the meso-level: something more than another term for individual action and yet less than the abstractions of structuration theories.

It is clear that such a discourse would apply to the whole universe of thinking about risk and safety, from nuclear deterrence theory and anti-meteorite missiles to playground equipment and fireproof pajamas. But rather than engage *that* universal discourse, I would like to limit myself here to middle-range theorizing and to studies of the operational and managerial dynamics of complex organizations or sociotechnical systems. For we have seen "safeness" in action, but we still find it hard to determine from whence it comes.

In this chapter, I first seek to identify ways to characterize safety as a positive characteristic of the relative success of organizations that have been operating complex technical systems under very demanding conditions. I then turn to the literature on risk and safety to illustrate the degree to which conventional approaches to safety cultures fail to capture the mythic and discursive dimensions of operational safety. Some salient descriptive properties that differentiate these organizations from others performing less well are identified. I then describe my attempts to use more constructivist approaches to move beyond measurable indicators and structural explanations, ending with the conclusion that we still do not have theories of collective agency adequate to describe how safety is constructed in real-world operational settings.

Safety as embodied myth and formalized ritual

What led to the pursuit of analogies to wellness was James Reason's work on comparing risk and safety to illness and health (Reason 1995). If one takes perceived risk as a constructed measure of the potential for error, safety stands in the same relationship to error and risk as health does to illness and the potential for being unwell. Reason further points out that health can be thought of as an emergent property inferred from a selection of physiological signs and lifestyle indicators. He then constructs an analogy with complex, hazardous systems, pointing out that assessing an organization's "safety health" resembles the medical case in that it requires sampling from a potentially large number of indicators. Although Reason's argument then proceeds in a more positivistic direction towards the development of separable indicators than the social constructivist perspective followed here, the analogy is nevertheless apt.

This sharpens the point about taking safety to be a positive construct and not just as the opposite of risk. Defining an organization as safe because it has a low rate of error or accident has the same limitations as defining health in terms of not being sick. When a person characterizes his/her medical state as one of being "healthy," s/he usually means far more than not ill, as is increasingly recognized by those who have adopted the term "wellness" rather than health. Other factors that may enter individuals' assessments of health include a calculation that they are not at the moment at any risk of being ill or that they have a cushion or margin of protection against illness, that is, they have a low potential for being unwell. In most cases, wellness goes beyond health in expressing a state of being with many dimensions, a state that is not entirely about the clinical state of the physical body but rather is a story about the relationship of the individual's mind and body with both real and perceived social and physical environments. The growing literature on wellness argues that perceptions such as being able to control one's environment promote being well, even when being well is measured as "good health" by objective indicators (Swartzberg and Margen 1998), an observation that has recently been emphasized by noting that even placebos can have positive effects (e.g. http://www.apa.org/releases/placebo.html).

Wellness is in some sense a story that one tells about one's self and one's relation to others and the world. Safety is in some sense a story that a group or organization tells about its self and its relation to its task environment. The origins of both do not lie solely with indicators or performance. Meyer and Rowan have noted that there is a correlation between on the one hand elaborate displays of confidence, satisfaction and good faith, and on the other hand the institutionalization of myth and ritual (Meyer and Rowan 1977). Moreover, these correlations are in most cases more than affirmations of institutionalized myths; they are the basis for a commitment to support such representation by collective action. Putting these insights together with the observations of the Berkeley HRO group, I argue that a collective commitment to safety is an institutionalized social construct. The organizations we observe not only perform well but also weave into their operations the kinds of rituals and stories that serve to orally transmit operational behavior, group culture and collective responsibility. And since these rituals and stories arise from the organization's perpetual reaffirmation of its commitment to safety not only as an institutionalized collective property but a constant process of re-enactment of its own stories, the result is in many ways a dynamic, interactive and therefore intersubjective process that can only be nurtured and maintained if the means and modes of social interaction are preserved and reinforced. But such processes do not, and cannot, proceed without human agency, and it is the nature and locus of that agency that has proved most elusive in the research and literature on risk and safety in technical operations.

Beyond normal accidents

When the Berkeley group on "high reliability organizations" (HROs) first set out in the mid-1980s to examine the role of operators in maintaining safety in a variety of complex, tightly coupled high-technology operations, there was very little guidance in the organizational literature. Perrow had just developed his "normal accident" framework to explain a set of rather dramatic accidents (or near-misses) in complex, tightly coupled technical systems (Perrow 1984). In contrast, the systems studied by our group seemed to have had a rate of error that was remarkably low compared to what the literature in general, and normal accident theory in particular, would lead one to expect. Keeping within the functionalist perspective, we therefore focused initially on structure and static indicators, seeking to learn how errors were avoided and risk thereby managed or controlled. What we found instead was an operational state that represented more than avoidance of risk or management of error. Those organizations characterized as "highly reliable organizations" all show a positive engagement with the construction of operational safety that extends beyond controlling or mitigating untoward or unexpected events and seeks instead to *anticipate* and *plan* for them. Unfortunately, the existing organizational literature was almost silent on this dimension of organizational behavior.

Normal accident theorists further warned that safety devices or other kinds of additional technical redundancy that were meant to control or reduce error can increase overall complexity and coupling, resulting in systems that are more prone to error than they were previously (Perrow 1994; Sagan 1993).

In our HRO work, operators of comparatively reliable systems were self-consciously aware of this problem; we found several instances in which operators opposed further introduction of supposedly error-reducing procedures or technologies for precisely those reasons (Rochlin 1986; Rochlin *et al.* 1987). The argument presented here takes that observation a step further, building on the argument from HRO research that some organizations possess interactive social characteristics that enable them to manage such complex systems remarkably well (La Porte 1996; Rochlin 1993), as well as the additional observation that we do not know enough about either the construction or the maintenance of such behavior to be confident about its resilience in the face of externally imposed changes to task design or environment (La Porte and Consolini 1991; Roberts 1993b).

Throughout our research, we observed again and again, in different contexts and manifestations, that maintenance of a high degree of operational safety depends on more than a set of observable rules or procedures, externally imposed training or management skills, or easily recognized behavioral scripts. While much of what the operators do can be formally described one step at a time, a great deal of how they operate, and – more importantly – how they operate safely, is "holistic" in the sense that it is a property of the interactions, rituals and myths within the social structure and beliefs of the entire organization or at least of a large segment of it (Weick and Roberts 1993; Weick *et al.* 1997).

The "safety culture" we observed in our work was not just a story the operators were telling to each other in search of common interpretations or common illusions, neither was it just an expression of organizational myths (Pidgeon 1991). It more closely resembled what Wiley has called "generic subjectivity," that is, the construction of a social culture that provides meaning and purpose to the system processes that people operate or manage (Wiley 1988). Swidler's notion of distinguishing strategies *in* action from strategies *for* action is also suggestive (Swidler 1986). Moreover, at least in the case of nuclear plants, air traffic control and aircraft carrier flight operations, this subjectivity is not just an intersubjective social construct within the group but "generically" subjective in that it reflexively incorporates group performance as part of the structure of the relationship between task, task environment, task group and organization. The more technically complex the system, the less possible it is to decompose the interlocking set of organizationally and technically framed social constructs and representations of the plant in search of the locus of operational safety.

Risk and safety

The instrumental literature on risk provides two sets of definitions of risk and safety. The more technical set, focusing largely on the causes and consequences of operational error, tends to define risk and safety instrumentally, in terms of minimizing errors and measurable consequences (Waterstone

1992). A second, newer thread of the literature is more socially and politically oriented, placing more emphasis on representation, perception and interpretation (Clarke and Short 1993; Jasanoff 1993; National Research Council 1996). My intention here is to consider safety even more broadly as a social construct – something that is not "out there," but really "in here," as a reflexive, dynamic and discursive property of operations and not a set of behavioral characteristics to be isolated, catalogued and measured.

Safety is not merely the absence of accident or the avoidance of error. For the operators of highly reliable systems such as air traffic control centers or well-performing nuclear plants, safe operation encompasses both compliance with external regulations and technical conditions and a positive state of operations that are reflexively expressed and mediated through human action. It is both agency and structure. But for the most part, structural analysis only addresses questions of risk avoidance or error prevention, similarly to most of the social science writing on risk and safety as well as the more technically and engineering-oriented literature.

The social science literature provides many different ways to express similar ideas, but none of them seem quite appropriate. There are means and methods ranging from frame analysis to representation to social construction and interpretation, depending upon the model that is used for analyzing the relationship between the actors, their equipment and the task at hand. But although each of the relevant literatures was suggestive, none were entirely satisfying as a mode of explanation for the Berkeley HRO group's observations, and none supplied a theory of agency sufficiently rich to provide an explanatory framework that fully incorporated the dynamic role of the operators in the process. Burns and Dietz, for example, begin their famous review of sociotechnical systems by asserting "We emphasize agency" (Burns and Dietz 1992). But when they analyze safety in nuclear power plant operations, they return to the normal accident model, which is explicitly structural and system-level. In the process, they shift from thinking about safety to analyzing sources of error and risk. In an attempt to transcend these limitations, several organization theorists have turned to ethnographic methods in their studies of technical work as "practice." But such studies still tend to site the observer in the organization rather than "of" it, as well as bring analytical structures to the field rather than adopting the classic grounded theory approach of letting them emerge from it (Ragin and Becker 1992; Strauss and Corbin 1990).

A great deal of the literature on risk or error that is done in these modes expresses to some extent the idea that safety is an active construct, but for the most part it treats the actors as "actors" in the sense of single-loop learning, i.e. acting within the constraints of internal models of organization and operation as well as exogenous social and technical circumstances. The Berkeley group found that this work usually failed to address how our "actors" were also "agents" in the sense of double-loop learning, i.e. capable not only of modifying both internal and external parameters but also of deconstructing and reconstructing these parameters through social interaction.

We observed that safety was difficult to specify even in the relatively simple domain of self-constructed expectations and demands. Most operators whom we studied reported that the chance of error or accident actually increased when those at the controls felt that they were in an "unsafe state." But even these operators were unable to provide a characterization of those conditions that would result in their making that type of judgment. They did not volunteer that the converse was true, i.e. that operations were more likely to be safe when operators were comfortable than when they were operating safely, even though that is the logical consequence of the previous, more negative statement. Indeed, operators tended to avoid the subject completely and to respond vaguely when asked about it. Such response is consistent with our finding that collective and interactive behavior is an essential part of the construction of safety in these organizations. This finding is not remarkable in an instrumental sense; many cognitive psychologists have found that interactive teamwork is essential for minimizing errors in the management of complex technical operations. But cognitive analysis of work pays insufficient attention to the perceptual and interpretive dimensions of safety, treating them primarily as "attitudes" or even "safety cultures" that can, in principle, be addressed by simple changes in task design or managerial behavior.

"Safety culture" as described in the management literature is usually a systemic, structural and static construct. In contrast, the culture of safety we in the Berkeley HRO group observed is a dynamic, intersubjectively constructed belief in the possibility of continued operational safety, instantiated by experience with anticipation of events that could have led to serious errors and complemented by the continuing expectation of future surprise. Rather than taking success as a basis for confidence, these operators, as well as their managers, maintain a self-conscious dialectic between collective learning from success and the deep belief that no learning can be taken to be exhaustive because the knowledge base for the complex and dangerous operations in question is inherently and permanently imperfect.

Some salient properties

How do these comparatively safe organizations differ from others for whom professions of being safe might be better characterized as expressions of over-confidence, poor judgment or excessive caution? In our view, the lack of self-promotion or aggressive self-confidence is in itself a useful gauge, as is the desire to attribute errors or mistakes to the entire organization as well as to human action. Nevertheless, the lack of a coherent theory of agency means that further efforts to determine the nature of the relationship between perceptions of and beliefs in safe operations and measurable indicators of performance still remain outside the scope of projective or prescriptive theorizing. For the present purpose, it is more appropriate to continue to be

descriptive, to focus attention on organizations that are broadly and generally considered to be safe, and to describe what we think are the most relevant,
unique and important safety-related observations that have come out of our
work rather than to try to generate a more comprehensive model.

Learning

In order to maintain safety in a dynamic and changing organizational setting
and environment, system operators and managers must continue to learn.
But by what standards can that learning be gauged? Considerable individual
learning takes place, and some of the organizations consider such learning
to be central to their performance. What is less clear without a theory to
determine the appropriate level of agency is whether individual learning
translates into organizational or institutional learning. In his analysis of risk
and social learning, Wynne has invoked a parallel with the reflexive and self-
conscious model of organizational learning developed by Argyris and Schön
(Argyris and Schön 1978, 1996), adding further that institutions which learn
do not exclude their own structure and social relations from the discourse
(Wynne 1992). Instead, Wynne asserts, they eschew narrow definitions of
efficiency or risk and embrace multiple rationalities:

> It is the connected dimensions, of self-exploration combined with devel-
> oping relations with others and their forms of knowledge and experience,
> which constitute the crucial *reflexive* and *interactive* learning. This has no
> preordained or guaranteed direction; indeed, it needs recognition of the
> *indeterminacy* of values, identities, and knowledges in order to be possible.

The essential indeterminacy of values, identities and knowledge is precisely
what characterizes the safety learning that we observed in our organizations,
with the added dimension that it is not just the organization as a whole
but differently positioned actor-groups who are learning, and that each learns
different things at different times from different signals sited or anchored
in different operational contexts. Also, the signals themselves are not always
obvious to even a trained observer. Because many of these organizations are
in a situation in which they cannot indulge in trial and error (precisely
because their first error might be their last trial), they learn not only from
near misses but also from situations that they evaluate as having had the
potential to evolve into near misses. Such learning is inherently reflexive,
since the ability to identify such situations is in itself part of the constructed
ambiance of safe operation.

Safety in action

In all of the systems studied by the Berkeley HRO group, the construction
of safety was an ongoing enterprise. In our work on aircraft carrier flight

operations, for example, we found a constant stream of talk taking place simultaneously over multiple communication links during flight operations, even when the level of operational activity was low (Rochlin *et al.* 1987). By comparing this exchange with other similar, if less dramatic, activities taking place elsewhere on the ship during critical operations, we found that the constant chatter was an important means for maintaining the integration of the collectivity and reassuring all participants of its status. Similar behavior was also seen in nuclear power plant control rooms (Rochlin and von Meier 1994). Air traffic control is radically different in that controllers interact more with pilots than with each other (La Porte and Consolini 1991). Nevertheless, Johan Sanne's analysis of extensive video and audio recording of conversations between controllers and pilots at a Scandinavian control center also displayed many interactions that were not part of the formal control process but were meant to nurture communication and cooperation between controllers and pilots (Sanne 1999). Their structure and content were shaped to a great extent by the recognition of all concerned parties that safety was indeed a collective property that "emerged" from the interaction between them.

Similar findings have long been reported in the relevant literature but perhaps have been insufficiently analyzed. The work of the Westinghouse group on nuclear power plant operations (Roth *et al.* 1992) and the concept of "social organizational safety" described by Janssens and colleagues (1989) point out the importance of the "spontaneous and continuous exchange of information relevant to *normal* functioning of the system in the team, in the social organization" (Janssens *et al.* 1989, my emphasis added). But these exchanges were examined primarily as conditions for instrumental performance and the avoidance of task saturation and representational ambiguities, without the added dimension of asking whether performance and capacity were interactively redetermined by beliefs that they could operate safely *even if the conditions of work or the bounding conditions of tasks had to be altered in the process.* In a sense, these analyses continued to depend implicitly on the idea of single-loop learning. In so doing, they missed the critical dimension of whether, and how, the capacity for double-loop learning was essential for maintaining the construct of safety – that is, whether and how agency was involved.

Collective action

Schulman has suggested a frame for understanding the importance of collective behavior in defining operational safety by noting a sharp contrast between organizations in which "hero stories" were prevalent and welcome and those in which such stories were rare and not encouraged (Schulman 1996). Organizations such as militaries and fire departments emphasize extraordinary individual performance and rapid response; hero stories become epics that define organizational culture while also being an effective means for organizational learning and the maintenance of cumulative knowledge. In contrast, in organizations within air traffic control and nuclear power, "hero" is a critique, describing an operator who will act with little regard for the collectivity. Safety is sought through collective action and through shared knowledge and responsibility. Organizational learning is formalized, and cumulative knowledge is transmitted not through legends or narratives but through double-loop learning – in the cases of air traffic control and nuclear power, specifically through recreating and modifying formal procedures.

The two differing representations of operator responsibility and operational safety lead to contrasting sets of social and individual behavior. The first set establishes a context in which responsibility for performance of non-routine tasks rests primarily with individuals, that is, potential heroes to be sought out and nurtured; what lends confidence in collective performance is the belief that any contingency will be met by one person or the other. This is very like the traditional medical definition of "health" as a state of the individual characterized by a lack of disease or other negative interventions. The second set of social and individual behavior rejects such an approach as dangerous, especially for those tasks that present the greatest potential for harm. Threats are to be identified, scoped and routinized; when unexpected events do occur, they must be responded to through collective action in which heroic individual initiative is a threat, not an opportunity. Safe operation is not just a positive outcome but a state of the system (like "wellness"), and the constructed narrative is one of organizational rather than individual performance.

This analysis helps us to understand one of the key findings of our HRO research: first, that reliable organizations tend to reward the reporting of error and the assuming of responsibility for error, and second, that even when human error does occur, the root cause is considered to lie with the organization as a whole rather than being (dis)placed on the erring task group or individual. Both of these findings are contrary to most of the literature on complex organizations. The second one is most remarkable in that it spans interventions ranging from taking overall responsibility for poor training or bad working conditions to attributing the bad performance of an unqualified individual in a responsible position to a collective failure in evaluation. There is no category of error or malfeasance that absolves the collectivity of responsibility for the outcome. Agency has become detached from individual actors and work groups and devolved on the whole in ways that are not yet fully understood.

Constructivist approaches

Analyzing the more subjective and social dimensions of the organizations that the Berkeley HRO group observed in the field proved difficult. Much of the existing theory and research experience remains grounded in structure-oriented, rational-action and task-centered approaches that are oriented towards finding *measurable indicators* of the interfaces and relationships between organizational characteristics, the performance of individual workers, the nature of their tasks and the properties of the equipment with which they work. This orientation extends not only to technical model building and organizational analysis, but also to human–technical interactions, psychological theories of framing and scenario-building and even cognitive analysis. One of the problems with these approaches is that important phenomena such as ritualization, seemingly non-functional displays and expressions of confidence that are not grounded in empirically determinable observables are all too often disregarded, or worse, trampled on or overwritten in an effort to impose theoretically argued strategies that are argued to be "functional" solely in terms of abstractly constructed models and formalized indicators.

In search of more constructivist approaches that would enable us to avoid the traps of embedded positivism and implicit structuralism (Weick 1999), the Berkeley HRO group examined a variety of narrative and post-positivist methods as potential tools for understanding safety as a property *in* the system rather than *of* it. At the same time, we needed to be careful to avoid what Bowker and Star have recently characterized as the imposition of values, prior beliefs and master narratives by an over-reliance on schemes of classification (Bowker and Star 1999). In so doing, we necessarily placed ourselves in opposition to those who seek grand narratives and deterministic outcomes as well as those who seek to construct general, formal models of organizational performance.

Cognitive ergonomics

In contrast to more organization-oriented theoretical approaches, cognitive ergonomics attempts to think directly about the operators and the machines they operate as integrated actors (Hollnagel 1997). But having a theory of action is not the same as having a theory of agency, and for the most part the operators remain operators of the machinery rather than with it. In the case of many increasingly complex and interactive systems, analysts and system engineers tasked with solving operational problems are often unable to localize and identify either the sources of interactive failures or the modes through which these failures interact. Instead, they focus on abstracted systemic, network and interactive models that do not replicate the experience of the operators and users of the real system, which in turn can destabilize the very processes (such as cognitive integration and cognitive trade-off) that operators use to build their own confidence and reliability.

The emerging field of cognitive systems engineering offers some promising approaches based on the recognition that no single theory has explanatory power across all the dimensions of the phenomenon of work in modern, highly interconnected information societies – let alone attempting to explain the *performance* of work. Recent analysis emphasizes that the operators do not work "in" context or "with" culture, but rather that operators, work, context and culture are all interactively and mutually constructed in sociotechnical systems. Hutchins, for example, has discussed the growth of cognition in complex systems as proceeding through a process of artifactual "anchoring," in which mental models and interpretations are bound up with the physical artifacts of the operational environment. This anchoring requires not only that the artifact or equipment is stable but also that its behavior is a reliable and predictable base upon which to build confidence and expertise (Hutchins 1995).

By focusing on linguistics and semantics rather than treating operators as information processing machines, studies adopting this approach implicitly (if not explicitly) also adopt notions of reciprocity and reflexivity. However, such a stance leaves the cognitive scientist in the jaws of the famous anthropologist's dilemma: how to construct a model of the system that is not just as much a projection of the analyst as it is a replica of the analyzed.

The early work of Jens Rasmussen and colleagues at the RISØ laboratory in Denmark was an exception in that it was more ethnographic and less formalized than traditional cognitive analysis (Rasmussen 1990; Rasmussen *et al.* 1991; Rasmussen *et al.* 1987). This work rested extensively on verbal protocols in which the operators were not asked to reflect introspectively on their work but merely to try to relate what they were doing in their tasks. This research has led to a growing appreciation of the importance of ethnographic observations for acquiring a basic understanding of cognition that has led *inter alia* to the development of approaches such as the context-framed approach of cognitive systems engineering (Flach and Kuperman 1998; Woods and Roth 1986).

Ethnographic approaches

But does ethnography really address the slippery problem of agency? Experienced ethnographers, particularly anthropologists, often assert that agency is an essential part of their findings (Bernard 1994). Agency is in turn central to modern social theory (see, for example, Giddens and Turner 1987). But the relative coherence of the definitions of agency in social theory is rapidly deconstructed in fieldwork studies, each of which seems to generate its own particular, context-specific definition – when it generates one at all. Also, instead of being guided by agency in the singular, complex organizations – with their highly specialized and differentiated technical subsystems that have equally complex and often indirect failure modes – seem to have multiple sites and forms of "agency," all of which may (or may not) be more (or less) essential to the construction of safety. How are we to perform the right kind of research on such organizations?

Few sociologists or political scientists are prepared to carry out full ethnographies on their domains of study, nor do they attempt to completely adopt the methods of deep ethnography. Instead, there are a variety of alternative approaches to the study of risk and safety in sociotechnical systems, exemplified by the extensive interviewing and immersion that characterizes Vaughan's work on how the structure and culture of NASA created the environment that framed the *Challenger* disaster (Vaughan 1996), the work of the Berkeley high-reliability group (Bourrier 1999; La Porte 1996; Roberts 1993a; Rochlin 1993; Weick and Roberts 1993) and context-dependency studies of nuclear power plants (Perin 1996). These studies might best be characterized as "semi" ethnographic: while they adopt the epistemological approach of ethnographies and accept the ontological uncertainty entailed, they usually try to keep the researcher at a greater analytic distance from the subjects at study than a modern ethnographer would be comfortable with (Van Maanen 1988, 1997). They rarely fully adopt the openly unstructured means and methods for gathering evidence in the field or the deep suspicion that even hypotheses generated in the field are reflexive constructs.

Participant observation is a promising method for field research, but for most of the complex organizations with which we are concerned – such as those in which human safety is at stake or those for which the technical demands require many years of apprenticeship and formal training – the demands and costs of becoming a real "participant" are simply too high for any but a few analysts to bear. It is not surprising that cases in which practitioners were analysts-in-training or later became analysts are rare; one exception is the domain of airplane flight, since at least a few analysts are or have been pilots as well. Assuming that it will be difficult to find analysts who are also trained practitioners in other domains, the only way to gain similar familiarity with the issues in other settings would be to extend, for example, conventional sociological analysis to include ethnographic studies and approaches (or at least semi-ethnographic ones).

Writing from a different perspective that has similar problems, namely the perspective of technology studies, Sanne argues that such a synthesis would move some meaningful distance toward gaining a better understanding of the situated knowledge – and equally situated problems – of operators in modern, complex, high-technology systems (Sanne 1999). The issue is not how people come to know what they know, but rather how the known is interactively constructed in the process of knowing. This process entails both learning as practice and practice as learning, both of which (at least in Sanne's case) are firmly anchored in the physical and technical artifacts that surround the task at hand and mediated and moderated by its interactive complexity.

Conclusion

The challenge in studying how risk and safety are constructed in complex organizations (or systems) is to gain a better understanding of the interactive dynamics of action and agency in these organizations, as well as the means by which action and agency are shaped and maintained. These dynamics arise as much from interpersonal and intergroup interaction as from more commonly studied interactions with regulators, officials and system designers. Because the interaction is a social construction anchored in cultural dynamics, there are wide variances in its manifestation even among, for example, similar technical plants in roughly similar settings. An even greater challenge for social scientists will be to identify rules that correlate constructions of operational safety with empirical observations of performance, so as to distinguish those organizations that construct an accurate representational framework of safety from those that only construct the representation – and not the safety.

Throughout our research, we in the Berkeley HRO group repeatedly observe in different contexts and manifestations, that "operational safety" cannot be captured as a set of rules or procedures, simple, or empirically observable properties, externally imposed training or management skills, or "deconstructable" cognitive or behavioral frames. While much of what operators in complex organizations or systems do can be empirically described and framed in positivistic terms, a great deal of how they operate – and, more importantly, how they operate safely – stems from and is deeply embedded in the interactions, rituals and myths of the social structure and beliefs of the entire organization (or at least of a large segment of it). Moreover, at least in the cases of highly reliable organizations such as nuclear power plants, air traffic control centers and aircraft carrier operations, safety is an emergent property not just at the individual level but also at the level of interactions among groups within the sociotechnical system as a whole.

However intersubjective socially constructed and collective these interactions, rituals and myths may be, they are central to understanding the role of agency in providing safety and, in turn, understanding the reliable operation of complex, potentially hazardous sociotechnical systems. To the extent that regulators, system designers and analysts continue to focus their attention on the avoidance of error and the control of risk, as well as to seek objective and positivistic indicators of performance, safety will continue to be marginalized as a residual property. This type of approach not only neglects the importance of expressed and perceived safety as a constitutive property of safe operation, but it may actually interfere with the means and processes by which such safety is created and maintained.

It is therefore important in practice, as well as in theory, to encourage continued and expanding research and inquiry into safety as an expression of interactions, myth and ritual, of agency as well as structure – and not just as a measurable, statistical property of organizations, a system for reporting work team errors, or a way of expressing the avoidance of consequential accidents. It will not be easy for social scientists to avoid the many pitfalls of multiple-method fieldwork (or multi-site ethnographies) and to walk a fine line between on the one hand the excessive formalism of traditional means of organizational research and on the other hand post-modern approaches that eschew relevance altogether. But at a time when almost all organizational theorizing is in a state of flux, and the safety of large, sociotechnical systems is questioned virtually everywhere, the possible gains are enormous.

References

Argyris, C. and Schön, D.A. (1978) *Organizational Learning : A Theory of Action Perspective*, Reading, MA: Addison-Wesley Pub. Co.

—— (1996) *Organizational Learning II: Theory, Method, and Practice*, Reading, MA: Addison-Wesley Pub. Co.

Beck, U. (1992) *Risk Society*, London: Sage.

—— (1996) "World Risk Society as Cosmopolitan Society – Ecological Questions in a Framework of Manufactured Uncertainties," *Theory Culture & Society* 13(4): 1–32.

Beck, U., Giddens, A. and Lash, S. (1994) *Reflexive Modernization: Politics, Tradition and Aesthetics in the Modern Social Order*, Stanford, CA: Stanford University Press.

Bernard, H.R. (1994) *Research Methods in Anthropology: Qualitative and Quantitative Approaches*, Thousand Oaks, CA: Sage.

Bourrier, M. (1999) "Constructing Organizational Reliability: The Problem of Embeddedness and Duality," in J. Misumi, B. Wilpert and R. Miller (eds) *Nuclear Safety: A Human Factors Perspective*, London and Philadelphia: Taylor & Francis.

Bowker, G.C. and Star, S.L. (1999) *Sorting Things Out: Classification and its Consequences*, Cambridge, MA: MIT Press.

Burns, T.R. and Dietz, T. (1992) "Technology, Sociotechnical Systems, Technological Development: An Evolutionary Perspective," in M. Dierkes and U. Hoffmann (eds) *New Technology at the Outset: Social Forces in the Shaping of Technological Innovations*, Frankfurt and New York: Campus/Westview.

Clarke, L. and Short, J.F. (1993) "Social Organization and Risk – Some Current Controversies," *Annual Review of Sociology* 19: 375–399.

Douglas, M.T. and Wildavsky, A.B. (1982) *Risk and Culture: An Essay on the Selection of Technical and Environmental Dangers*, Berkeley and Los Angeles: University of California Press.

Edlin, G., Golanty, E. and McCormack Brown, K. (eds) (1996) *Health and Wellness*, Sudbury, MA: Jones and Bartlett.

Flach, J. and Kuperman, G.G. (1998) *Victory by Design: War, Information, and Cognitive Systems Engineering*, Wright-Patterson AFB, OH: United States Air Force Research Laboratory.

Freudenberg, W.R. (1988) "Perceived Risk, Real Risk: Social Science and the Art of Probabilistic Risk Assessment," *Science* 242: 44–49.

—— (1992) "Heuristics, Biases, and the Not-so-General Publics: Expertise and Error in the Assessment of Risks," in S. Krimsky and D. Golding (eds) *Social Theories of Risk*, Westport, CT: Praeger.

Giddens, A. and Turner, J.H. (eds) (1987) *Social Theory Today*, Stanford, CA: Stanford University Press.

Hollnagel, E. (1997) "Cognitive Ergonomics: It's All in the Mind," *Ergonomics* 40(10): 1170–1182.

Hutchins, E. (1995) *Cognition in the Wild*, Cambridge, MA: MIT Press.

Janssens, L., Grotenhus, H., Michels, H. and Verhaegen, P. (1989) "Social Organizational Determinants of Safety in Nuclear Power Plants: Operator Training in the Management of Unforeseen Events," *Journal of Organizational Accidents* 11(2): 121–129.

Jasanoff, S. (1993) "Bridging the Two Cultures of Risk Analysis," *Risk Analysis* 13(2): 123–129.

Kasperson, R.E. (1992) "The Social Amplification of Risk: Progress in Developing an Integrative Framework," in S. Krimsky and D. Golding (eds) *Social Theories of Risk*, Westport, CT: Praeger.

La Porte, T.R. (1996) "High Reliability Organizations: Unlikely, Demanding, and at Risk," *Journal of Contingencies and Crisis Management* 4(2): 60–71.

La Porte, T.R. and Consolini, P.M. (1991) "Working in Practice but not in Theory: Theoretical Challenges of 'High-reliability Organizations'," *Journal of Public Administration Research and Theory* 1(1): 19–47.

Luhmann, N. (1993) *Risk: A Sociological Theory*, New York: de Gruyter.

Meyer, J.W. and Rowan, B. (1977) "Institutionalized organizations: Formal Structure as Myth and Ceremony," *American Journal of Sociology* 83(2): 340–363.

National Research Council (1996) *Understanding Risk: Informing Decisions in a Democratic Society*, Washington, DC: National Academy Press.

Perin, C. (1996) "Organizations as Context: Implications for Safety Science and Practice," *Industrial & Environmental Crisis Quarterly* 9(2): 152–174.

Perrow, C. (1984) *Normal Accidents: Living with High-Risk Technologies*, New York: Basic Books.

—— (1994) "The Limits of Safety: The Enhancement of a Theory of Accidents," *Journal of Contingencies and Crisis Management* 2(4): 212–220.

Pidgeon, N.F. (1991) "Safety Culture and Risk Management in Organizations," *Journal of Cross-Cultural Psychology* 22(1): 129–140.

Ragin, C.C. and Becker, H.S. (1992) *What is a Case? Exploring the Foundations of Social Inquiry*, Cambridge and New York: Cambridge University Press.

Ralph, R. (ed.) (1988) *Probabilistic Risk Assessment in the Nuclear Power Industry: Fundamentals and Applications*, New York: Pergamon Press.

Rasmussen, J. (1990) "Human Error and the Problem of Causality in the Analysis of Accidents," *Philosophical Transactions of the Royal Society of London* B 327: 449–462.

Rasmussen, J., Anderson, H.B. and Bernsen, N.O. (eds) (1991) *Human–Computer Interaction*, London: Erlbaum.

Rasmussen, J., Duncan, K. and Leplat, J. (1987) *New Technology and Human Error*, New York: John Wiley & Sons.

Rayner, S. (1992) "Cultural Theory and Risk Analysis," in S. Krimsky and D. Golding (eds) *Social Theories of Risk*, Westport, CT: Praeger.

Reason, J. (1995) "A Systems Approach to Organizational Error," *Ergonomics* 39(8): 1708–1721.

Renn, O. (1992) "Concepts of Risk: A Classification," in S. Krimsky and D. Golding (eds) *Social Theories of Risk*, Westport, CT: Praeger.

Roberts, K.H. (1993a) "Some Aspects of Organizational Cultures and Strategies to Manage Them in Reliability Enhancing Organizations," *Journal of Managerial Issues* 5(2): 165–181.

—— (ed.) (1993b) *New Challenges to Understanding Organizations*, New York: Macmillan.

Rochlin, G.I. (1986) "High-reliability Organizations and Technical Change: Some Ethical Problems and Dilemmas," *IEEE Technology and Society*, September 1986: 3–9.

—— (1993) "Defining High-reliability Organizations in Practice: A Taxonomic Prolegomenon," in K.H. Roberts (ed.) *New Challenges to Understanding Organizations*, New York: Macmillan.

—— (1999) "The Social Construction of Safety," in J. Misumi, B. Wilpert and R. Miller (eds) *Nuclear Safety: A Human Factors Perspective*, London and Philadelphia: Taylor & Francis.

Rochlin, G.I., La Porte, T.R. and Roberts, K.H. (1987) "The Self-designing High-reliability Organization: Aircraft Carrier Flight Operations at Sea," *Naval War College Review* 40(4): 76–90.

Rochlin, G.I. and von Meier, A. (1994) "Nuclear Power Operations: A Cross-cultural Perspective," *Annual Review of Energy and the Environment* 19: 153–187.

Roth, E.M., Mumaw, R.J. and Stubler, W.F. (1992) *Human Factors Evaluation Issues for Advanced Control Rooms: A Research Agenda*, New York: Institute of Electrical and Electronics Engineers.

Sagan, S.D. (1993) *The Limits of Safety: Organizations, Accidents, and Nuclear Weapons*, Princeton: Princeton University Press.

Sanne, J.M. (1999) *Creating Safety in Air Traffic Control*, Lund: Arkiv Förlag.

Schulman, P.R. (1996) "Heroes, Organizations, and High Reliability," *Journal of Contingencies and Crisis Management* 4(2): 72–82.

Strauss, A.L. and Corbin, J.M. (1990) *Basics of Qualitative Research: Grounded Theory Procedures and Techniques*, Newbury Park, CA: Sage Publications.

Swartzberg, J.E. and Margen, S. (1998) *The UC Berkeley Wellness Self-care Handbook: The Everyday Guide to Prevention & Home Remedies*, New York: Rebus (distributed by Random House).

Swidler, A. (1986) "Culture in Action: Symbols and Strategies," *American Sociology Review* 51: 273–286.

Van Maanen, J. (1988) *Tales of the Field: On Writing Ethnography*, Chicago: University of Chicago Press.

—— (1997) "The Asshole," in V.E. Kappeler (ed.) *The Police and Society: Touchstone Readings,* Prospect Heights, IL: Waveland Press.

Vaughan, D. (1996) *The Challenger Launch Decision: Risky Technology, Culture, and Deviance at NASA*, Chicago: University of Chicago Press.

Waterstone, M. (1992) *Risk and Society: The Interaction of Science, Technology, and Public Policy*, Dordrecht: Kluwer Academic Publishers.

Weick, K.E. (1999) "Theory Construction as Disciplined Reflexivity: Tradeoffs in the 90s," *Academy of Management Review* 24(4): 797–806.

Weick, K.E. and Roberts, K.H. (1993) "Collective Mind in Organizations: Heedful Interrelating on Flight Decks," *Administrative Science Quarterly* 38(3): 357–381.

Weick, K.E., Sutcliffe, K.M. and Obstfeld, D. (1997) *Organizing for High Reliability: The Suppression of Inertia Through Mindful Attention*, University of Michigan Business School.

Wiley, N. (1988) "The Micro–Macro Problem in Sociological Theory," *Sociological Theory* 6: 254–261.

Woods, D.D. and Roth, E.M. (1986) *The Role of Cognitive Modeling in Nuclear Power Plant Personnel Activities*, Washington, DC: US Nuclear Regulatory Commission.

Wynne, B. (1992) "Risk and Social Learning: Reification to Engagement," in S. Krimsky and D. Golding (eds) *Social Theories of Risk*, Westport, CT: Praeger.

7 Creating trust and achieving safety in air traffic control

Johan M. Sanne

The controllers give the crew information but cannot give them orders; they do not have any responsibility. The controllers show confidence in the pilot's ability to handle a situation. It is a question of mutual trust: as a pilot, you must trust the controllers, for example if you are in the clouds.

(Peter, flight captain, quoted in Sanne 1999: 218)

Risk is inherent to the ordinary operation of the many sociotechnical systems on which we rely in our daily lives. Sociotechnical systems – such as air transportation systems, energy systems and telecommunication systems – consist of technology, actors and organizations that are inextricably interwoven in a "seamless web" of interacting and mutually shaping components (Hughes 1986). Despite this complexity, statistics show that quite a few of the systems that involve the greatest hazards also show the lowest probabilities for fatal accidents (Sanne 1999: 13). How can this success be explained? How are technology, actors and organizations interwoven in the processes by which risks are handled and safety is achieved? This chapter will address these questions from the perspective of air traffic control, focusing on how mutual trust is created and safety is constructed by the involved actors.

Air traffic control is an example of a system that is characterized by both major hazards and a low probability of accidents. Flying involves many risks, such as the risk of collision with the ground (or other aircraft) and risks due to weather or technical troubles. Collisions that involve high-speed aircraft, in particular, typically entail many hazards. However, as the captain quoted above argues, these hazards are at least partly dealt with by building and maintaining mutual trust among air traffic controllers and pilots. When pilots fly in clouds, they cannot see anything and must rely totally on the air traffic controllers' guidance to avoid accidents. Conversely, controllers have to trust that flight crews will do their best to comply with the guidance that the controllers provide. Without trust in the nature of flight crews' future actions, the controllers would not be able to predict the traffic flow, which would make their task very complex and difficult.

The risks of flying are not restricted to the risks involved in managing individual aircraft. They are also related to certain characteristics of the air traffic control system, such as a high complexity and tight coupling between system components (Perrow 1984). Sociotechnical systems such as air traffic control systems are also characterized by a high degree of *specialization of tasks* among several actors and organizations (such as flight crews, air traffic controllers, weather forecast services and maintenance crews) as well as a high degree of *interdependence* among them. Interdependence creates a need for mutual understanding and coordination among people with different knowledge, different training and sometimes conflicting interests. The people who work in the various organizations must cooperate in order to produce a safe traffic flow, and such joint actions must often be achieved across distance and professional boundaries. The various actors depend on each other to coordinate and carry out specific actions in specific ways and at specific times (Sanne 1999).

Trust in one's colleagues is a salient part of the constitution of the air traffic control team. Gras *et al.* (1994) emphasize the importance of air controllers as a social group. Gras *et al.* argue that the risks that arise in air traffic control work and the consequences of individual mistakes or mishaps are so great that the individual controller cannot take sole responsibility for the risks pertaining to her/his tasks. Instead, the controllers *must jointly assume responsibility* for the risks that are imposed on them. Similarly, Randall (1994) argues that there is a need for a "corporate approach to failure" in air traffic control. For example, air traffic controllers' interventions to avoid mid-air collisions have often been the result of the controllers' constant attention to one another's work. In a system involving potential risks, participants need to have trust in each other's mutual support, in each other's competence and in their shared commitment to the collective achievement of safety (Suchman 1993).

In this chapter I will argue that the risks of air traffic control systems are handled through the production and acknowledgment of trust among the people involved. I will analyze why trust is needed, what kind of trust is involved in air traffic control operations, how it is produced and acknowledged and how it contributes to overcoming the risks that arise. I will argue that the structural properties of high complexity and tight technical coupling in air traffic control are overcome through "common orientation" and "tight social couplings" that provide the means for creating safety through trust (see, for example, Weick and Roberts 1993). The empirical material in the chapter is based on extensive fieldwork, primarily at the Terminal Control Center at Stockholm-Arlanda airport from 1996 to 1998 (Sanne 1999).[1]

Means and resources for creating trust

In air traffic control systems, trust serves two important functions: the need to facilitate cooperation and the need to achieve collective risk handling. The need for both of these functions stem from a certain degree of complexity and tight coupling of such systems. Interdependence and the potential of fatal, but inevitable, individual errors are structural properties of the air traffic control system that contribute to the creation of risks.

By what means or resources is trust built and safety achieved in the ongoing operation of such systems? It is clear that various technologies at hand are essential resources for achieving smooth operation of the system and safe handling of potential incidents. Besides these technical means, two other important resources for creating trust and safety are (1) a common orientation to the task, and (2) redundancy. These means or resources provide the basis for the cognitive and social integrative work that controllers and flight crews perform.

Common orientation to the task at hand is essential to achieving smooth cooperation among air traffic controllers and to the safe handling of potential incidents (Sanne 1999). Common orientation refers to the interrelated attention of different participants within the system. While performing individual tasks, the participants simultaneously attend to the overall state of the system, to the overall goals and to others' actions to this end (Sanne 1999). Common orientation involves understanding one's own individual contribution to the state of the system and to how things should be (that is, to a normative ordering principle that governs the work). Because there is a need to keep track of action and make sense of what is happening, the participants must know the system as a whole and their role in it. Therefore, common orientation refers to both a common understanding of the situation at hand and to commitment to the team and to the task.

Common orientation[2] has been described in ethnographic studies of teamwork in sites such as chemical plants, air traffic control towers, subway control centers and ship navigation centers (Bergman 1995; Gras *et al.* 1994; Heath and Luff 1992; Hirschhorn 1984; Hutchins 1995; Suchman 1993). In these systems, there is a widespread division of labor and thus a corresponding need to coordinate activities among the people involved; these people include participants within the control room (or cockpit) as well as those outside.

Mutual support is also provided through *redundancy*. As Hutchins (1995) argues, redundancy is produced in a team in which participants have overlapping knowledge of each other's tasks and will be able to correct mistakes made by individuals. There is redundancy both in the distribution of knowledge of individual tasks and in information, since the operators can overhear each other's talk and draw conclusions from it (Hutchins 1995). However, to make redundancy work as intended, it is necessary that the individuals in the team be attentive not only to their own tasks but also to each other's situation and performance. They must have the ability, time and motivation to achieve such attentiveness.

In a study of a chemical plant, Bergman (1995) observed that each worker had an area for which she or he was responsible during a specific shift. All of the workers were competent to handle a number of areas and they rotated between different areas from shift to shift. They were also attentive, however, to what happened in other areas outside the one they were currently responsible for, thereby providing back-up for each other. As one of them told the ethnographer: "In reality we are working on one and a half areas: you are chiefly responsible for one area but you are constantly prepared to intervene in other places" (Bergman 1995: 289). Due to the potential dangers and the potential consequences of different actions, the operators of such plants also need to be involved and committed to their work, to be "present" in the process. They have to be attentive to what is going on in terms of cues about something that might become problematic (Bergman 1995; Hirschhorn 1984).

To achieve trust, however, it is not enough that work be based on common orientation and that mutual support provide safety through redundancy. Those involved must also have good reasons to believe that their colleagues' work is informed by a common orientation and mutual support. That is, common orientation and the existence of redundancy has to be demonstrated and acknowledged. This is possible through, first, making one's work visible and, second, performing certain communicative practices that are an integral part of everyday work.

Visibility of colleagues' work makes it possible to learn what to expect and to detect possible errors. Through recurring patterns of behavior and a shared understanding of the system, others' work is made visible, intelligible and "readable," thereby providing grounds to know what to expect from them (Harper and Hughes 1993). Gras *et al.* (1994) point to the fact that the potential for interventions depends on the visibility of controllers' work and the ability for colleagues to keep on eye on one another. Controllers must be able to see their colleagues' decisions, they must understand their situation, and they must have time to do so (Gras *et al.* 1994; Randall 1994).

An example of *communicative practices* that serve to demonstrate and acknowledge common orientation and redundancy is controller–pilot talk, which can be analyzed within the frameworks of conversation analysis and theories of "institutional dialogue." Drew and Sorjonen (1997) point out that certain forms of "institutional dialogues," such as those used to achieve situated, collaborative practices in the workplace, are different from many of the dialogues that are commonly analyzed in communication studies. These collaborative practices include those whereby:

> colleagues manage and accomplish their work activities collaboratively – sometimes "co-constructing" a joint activity across distributed systems, some of which may involve technologically mediated communication systems, including those linking participants which are not physically co-present (for example, interactions between aircraft pilots and air traffic controllers).

> (Drew and Sorjonen 1997: 111–112)

The participants construct joint activities (such as avoiding the collision of two aircraft) through their interaction, even though they are distributed across a system and physically separated.

In the following sections I will analyze how cooperation is facilitated and how collective risk handling is achieved through processes of creating trust. In both cases, it is shown how visibility and communicative practices are integral to producing trust. The conditions for creating trust are described, and both successful and failed processes (and their causes) are analyzed. In both sections I stress the essential role of producing displays and accounts of trustworthiness through an informed use of the technologies at hand. Next, I discuss the implications this practice has for a successful introduction of new technologies.

Easing cooperation: through common orientation

The Pilot's Prayer

Oh controller who sits in tower
Hallowed be thy sector
Thy traffic come, thy instructions be done
On the ground as they are in the air
Give us this day our radar vectors
And forgive us our ABC incursions
As we forgive those who cut us off on final
And lead us not into adverse weather,
But deliver us our clearances.

Roger

The Pilot's Prayer can be found in several versions on the Internet. It tells us something about the relationship between controllers and pilots and the role played by radio talk in their establishing representations of each other and each other's trustworthiness.

First of all, we get a glimpse of how the work conditions of pilots and controllers differ. The pilot in the aircraft is exposed to adverse weather, whereas the controller in the control room is sheltered from it. The crew is enclosed in the cockpit and has a linear conception of the flight. To the controllers, the crew represents only one part in an overall situation with other aircraft involved; that is why the flight crews are sometimes "cut off on final." The aircraft passes from sector to sector and the controllers are called upon anonymously by their currently assigned control frequency ("hallowed be thy sector"), not in person. To the aircraft crew, the controllers are interchangeable parts of a system. Similarly, to the controllers, the crews are anonymous and are called upon by the call sign assigned to the aircraft or flight.

Second, *The Pilot's Prayer* indicates that the division of labor between controllers and flight crews carries with it a mutual, but asymmetrical, dependency. The poem tells us about the pilots' dependence on the controllers in several respects. Pilots must have the controllers' clearance to perform any trajectory in controlled airspace. They sometimes need to be given "radar vectors," i.e. compass headings, instead of using radio beacons when navigating (when controllers need to handle increased traffic efficiently). Pilots also sometimes have to accept that they are "cut off on final," i.e. when they have to make a go-around. This may be because the controllers have information about the current weather situation and use this command to avoid adverse weather. The poem also briefly mentions the controllers' dependence on the pilots' acceptance in order to fulfill their tasks: "And forgive us our ABC incursions."

Finally, the poem hints at the importance of the radio medium in determining how the relationship between controllers and flight crews evolves. Radio talk is the radar controllers' only means of influencing the aircraft's movement in the sky and vice versa; all other technologies are only informational. The medium imposes constraints on what can be communicated and achieved in the interaction between controllers and flight crews. But it also entails possibilities for the creation of trust. Each controller in operative control is assigned one radio frequency, on which she or he alone can talk for the assigned time. Everything the controller says is sent out and received by all aircraft that have tuned in to that frequency. The radio equipment is designed in such a way that when someone talks, no one else can talk at the same time on the same frequency; there is no possibility for simultaneous talk. All talk on radio is therefore public, both to all pilots listening to the frequency and to all controllers that listen to their colleagues' talk as it is heard "in the air" in the control room. This phenomenon is called the "party-line" character of radio talk (Eurocontrol 1996). Although the conversation between a controller and a particular flight crew concerns the handling of their particular aircraft, other crews can, and often do, listen to that talk.

The public nature of radio talk means that actions taken by any participant will be known to others who are listening to the radio. What the controller says to particular flight crews will influence other crews' opinion of him or her and of the particular air traffic control center. What other controllers hear from him or her "in the air" in the control room is also accounted for as evidence of that particular controller's trustworthiness. Radio talk is thus taken as an indication of proficiency, service-mindedness, trustworthiness, etc. (Randall 1994). The controllers can use this feature of radio talk by orienting themselves to the needs and interests of the flight crews and by showing compliance and cooperation with their particular needs.

Talk between controllers and pilots is highly regulated. What can be said is restricted, as are the specifics of "turn-taking" and the specific form that the expressions may take. The need for accuracy and the need for efficiency have structured the design of controller–pilot phraseology (Morrow *et al.* 1994). In actual controller–pilot talk, however, greetings and farewells are common. So are expressions of gratitude. Management in air traffic control has tried for a long time to eliminate these conventions of ordinary talk from controller–pilot dialogue, but so far without success. The controllers say "it is nice" to express these conventions. They like radio talk because it gives them a "personal contact" with flight crews. However, this is not the only or most important reason why controllers would find it unfortunate if they had to cease radio talk. It is more than just a question of the "politeness" or pleasantness of having a personal contact with flight crews, although this might also be important for achieving social cohesion within the system. As we shall see below, controllers and flight crews use these expressions as devices to achieve smooth interaction and accomplish specific goals. Such communicative strategies help achieve collaboration through making one's work visible and "readable" to the others.

Below, I will show two examples of how controllers display an orientation to the others' (both controllers' and flight crews') needs and interests while negotiating the specifics of individual flight routes. The first example concerns a "gift" that the controllers volunteer to give to the flight crew. They will make their handling of the traffic fit with individual flight crews' interests without being asked to do so:

Martin, controller at the Terminal Control, is currently working at *Departure-North*, talking to Braathens 455, approaching from Oslo:

Martin (Departure-North)	Braathens four five five, how many miles do you expect for descent?
Pilot (Braathens 455)	Twenty-five miles, Braathens four five five.

Martin produces a vector on the radar screen from a point near the runway that stretches 25 miles towards the symbol of Braathens 455.

Martin (Departure-North)	Braathens four five five, you will shortly have a left turn.

Martin explains to me that "You give them information about track miles (what is left of the flight) so that they can know when it is time to descend. Then the crew can plan their descent in advance, make it acceptable to the passengers and fuel-efficient." Martin can turn the aircraft wider if they need a longer distance to plan their descent.

In this case, common orientation was demonstrated and acknowledged without problems. In my second example, the creation of controller–pilot trust is part of the accomplishment of a separation between two aircraft. Here, it is the controller who asks for a "favor."

SAS 403 has departed from Stockholm's Arlanda airport and contacts the terminal control, but the crew does not get the altitude clearance it expects to have. Instead the controller issues a different clearance: "Maintain six thousand feet due to traffic." After a minute or so when SAS 403 has passed beneath an aircraft flying in the opposite direction, the controller gets back to the crew: "Scandinavian four zero three, cleared for flight level one seven zero, thanks for your cooperation." The deviation from the usual climb pattern causes the aircraft to interrupt the climb before it is established at cruise level: this increases the fuel consumption and is thus not in the interest of the company, i.e. the controller's customer.

In this case, the controller knew she had asked for a favor, namely requesting that the crew abstain from the usual flight route, the default flight clearance. To achieve this she explained the reason for the deviation to the default procedures ("due to traffic")[3] thereby presenting legitimate reasons for her request. She also acknowledged the help that the flight crew offered in order to help her separate the traffic. The changed flight route was clearly not in the flight crew's interest, but since the controller informed them of the reasons for the changed route, it was seen as appropriate. Her acknowledgment shows her orientation to the flight crew's interest and makes her actions seem not only appropriate, but also proficient and thereby trustworthy. However, her proficiency had to be made explicit, as it could not be demonstrated otherwise.

Next, I will provide an example in which there is no explicit account of proficiency to the flight crew. They therefore argue with the controller, based on their evaluation of the situation at hand. Their evaluation can be misleading, though, as can be seen in the following:

I am seated with Peter and Sven in the cockpit. We are at Stockholm's Arlanda airport, about to leave for Linköping (about 250 kilometers south). We are number two in line, waiting to taxi to the holding point from where to take-off. There is no aircraft at the holding point. The crew in the aircraft in front of us calls the tower controller and asks for permission to line up. The controller answered: "Negative, we have one landing ahead." To Peter and me, Sven comments: "Strange, Finnair [the aircraft just before] got it fast." Shortly thereafter a small propeller aircraft lands on the runway.

The crew in the aircraft in front of us could not understand why there was no aircraft waiting for take-off. The preceding aircraft (the Finnair) was long gone and it was therefore time for the aircraft in front of us to line up. The crew of that aircraft might have suspected that the controllers were too slow or inattentive; their request to take off indicated that they did not consider the controller's actions to be appropriate in this situation. But the controller's answer revealed that there was another scheme guiding his actions. The controller was landing an aircraft on the deviant runway (i.e. on the runway currently assigned for departures). Such an action is not immediately understood and accepted by flight crews, especially when it delays them personally. The flight crews are, as *The Pilot's Prayer* indicates, enclosed in the cockpit and have a limited view of the situation at hand.

There are stories circling around the world of situations in which controllers have lost confidence in the proficiency or appropriateness of flight crews' actions:[4]

> A DC-10 had an exceedingly long rollout after landing with his approach speed a little high.

> San Jose Tower: "American 751 heavy, turn right at the end of the runway, if able. If not able, take the Guadalupe exit off Highway 101 and make a right at the light to return to the airport."

In the way the story is told, it is the word "exceedingly" that conveys the moral of the story: it is the flight crew's inability to keep up with the expected rollout length (meaning getting off the runway after landing, thereby making room for the next aircraft to land) that provokes this "reading" of the flight crew's trustworthiness.

These few examples indicate a general situation. In order to achieve cooperation with flight crews while negotiating individual flight routes, controllers need to establish trust in their proficiency. Flight crews are dependent upon controllers' actions, and the two groups are anonymous to each other. Through talk and the mutual exchange of "gifts" or favors, the groups establish a common orientation and sustain social bonds between them. Trust has to be continuously negotiated, however. It may turn into mistrust should flight crews or controllers be confronted with inappropriate actions from the other side.

Handling risk through mutual support

As Randall has observed (1994), it is important for controllers to maintain a sense of collegiality. This collegiality entails proving one's own accountability and trustability, acknowledging that of others and showing that one always considers the needs and interests of colleagues. Trust among controllers is thus created within the community in at least three different ways.

First, trust is produced during negotiations between controllers about the handling of specific aircraft. A "coordination" in air traffic control refers to a process of negotiating the terms for the hand-over of the responsibility for the aircraft between controllers with responsibility for different areas. "Coordinations" are often not achieved explicitly through talk. Most of the time, controllers keep to their institutional roles in their interactions in order to achieve coordination. But interaction is also important in creating and maintaining interpersonal relations and in making decisions about "coordinations" visible to colleagues. "Coordinations" are therefore important not only to the achievement of practical coordinative work, but they also help to establish and display the accountability of others, as illustrated by the following account:

It is peak hour at Arlanda airport outside Stockholm. Controller Johanna is working at the area control, sector 5. She is talking to the crew of SAS 1130. The aircraft is carrying out a holding procedure at the radio beacon at Trosa (located at the entry to the Stockholm terminal area), but she is about to release it for approach into the terminal area. Johanna calls Oskar at the *Approach-Coordinator* workstation:

Johanna (R5)   SAS 1130 comes directly to Tebby
Oskar (Approach-Coordinator) Directly to Tebby

Oskar turns to Marianne at Arrival-East and points to her radar screen:

Oskar (Approach-Coordinator) It goes directly to Tebby
Marianne (Arrival-East)   OK

Johanna suggested that she should give SAS 1130 clearance for a shorter approach route directly to Tebby from Trosa instead of taking the longer default route. She recognized that if SAS 1130 followed the default route, the distance from the preceding aircraft would be larger than necessary, thereby reducing the number of aircraft landed per hour. However, with the flight route that she proposes the distance is reduced, which means that were will not be an "unnecessary delay." Her action is thus favorable to the terminal control, helping them to increase their productivity.

In the "coordination" described above, Johanna attends to the situation in the terminal area and suggests a procedure that is finely attuned to the interests of the terminal control team. She thereby not only achieves a safe landing but also demonstrates her proficiency, trustworthiness and collegiality within the community.

A second way in which trustworthiness is created can be seen in colleagues' everyday work, primarily in the image on the radar screen, which can be accounted for as the result and intention of their work (Harper and Hughes 1993; Randall 1994). Controllers can understand from the others' way of handling their traffic that their own interests are taken care of, without their needing to ask for such consideration. Their colleagues' experience and familiarity with the traffic situation and the demands that this places on individuals provide the grounds for such solidaristic acts. For example, controllers are sensitive to circumstances in which they should not disturb colleagues with, for example, "coordinations." I asked Anna during her trainee period whether it was difficult to understand what was happening around her, but she denied that this was the case. She could easily see if someone was busy:

> You can see if someone is busy, both on the terminal control and at the tower [from the image on the closed circuit TV]. In these cases you don't call them: you wouldn't like someone to call you if you were in that situation yourself. You don't call them, you make up something else.

If the colleague was busy, Anna tried to find a solution that did not force the colleague to coordinate with her. In this case, Anna supported the colleague indirectly by deciding not to coordinate and thereby to avoid disturbing her colleague.

There are of course instances when controllers think they have reasons to mistrust the competence of colleagues when "reading" their actions. In Sanne (1999: 192), controller Ulrik at the Stockholm terminal control had organized holding procedures using a "list," a method that involves organizing work according to a different mode of division of labor from the usual. The list helps the terminal controllers organize peak traffic more efficiently than by ordinary means. But it places an extra workload on their colleagues at Area Control and is therefore not welcome there, especially in moments when they cannot see any reasons for it.

One morning I am sitting in the "day-room," listening to the controllers. Maria, who is a controller at one air traffic department, is critical of Ulrik, who had been working at another department. Ulrik has first decided "going into the list," i.e. setting up the list, and then canceled it ten minutes later. Holding involves a lot of work for Area Control: in this case, they had called for three additional controllers to support the radar controllers. But "when we finally got some structure into our work, we left it [the list]," Maria explains. She and the other controllers at Area Control feel that the work is "particularly troublesome today." Coordination in the mode suggested by Ulrik takes more time and disrupts the flow. The controllers at Area Control feel that Ulrik had set up the list unnecessarily.

Ulrik set up the list since he thought it would be necessary, but in hindsight it was not, and his proficiency was called into question. Work did not break down due to mistrust, but the controllers had to discuss his decision later in order to restore understanding between the two air traffic control departments.

A third way to produce trust is through the monitoring and support of each other's work. Air traffic controllers have overlapping competencies and they routinely monitor each other's work to check for mistakes, omissions or errors. They support each other by recognizing situations, detecting deviations, defining problems and helping each other to handle these problems individually or by working together. As a team, they achieve safe and efficient traffic handling. This kind of monitoring and mutual support is, as Gras *et al.* (1994) and Randall (1994) argue as well, an absolute necessity. Errors are inevitable but may have catastrophic consequences. Mutual monitoring is therefore not regarded by the controllers as an indication of mistrust. On the contrary, controllers need to know that their colleagues routinely keep an eye on their work so that they can feel at ease when handling their tasks.

Controllers monitor colleagues' work either while performing their own tasks or while standing behind the others, watching the radar screen and sometimes overhearing their radio talk. This overlap is a form of redundancy in the organization. Mutual monitoring uncovers phenomena that the responsible controller may have missed, and it is performed for experienced and for more recently licensed colleagues alike. Because they have less experience, the latter are generally given more help than others, but in certain situations any controller will appreciate this kind of support, for example in cases of snow clearings of the runways, or during thunderstorms.

Monitoring and support are carried out both spontaneously in an informal and non-regulated form and within institutionalized forms. I was interested in the way such support is organized or activated and what it involved. Controllers rely on others to help them carry out actions safely. This support may be given either as guidance from a position of better knowledge of the situation at hand or as a kind of double check on the appropriateness of these actions, done by colleagues that have the same kind of information. For example, controllers seated adjacent to each other may listen to each other's radio talk or watch the radar image and detect possible errors or problems that arise in their colleagues' work.

As shown in Sanne (1999, 2001a), the experienced controller is able to recognize complex situations immediately. This effortless, routinized sense making means that experienced controllers have a redundant capacity in relation to their current individual task; they have time left for other tasks, especially in low-volume traffic situations. Controller Nils tells me that "as a controller you are curious." If he has some time to spare, he turns on the "quick look," a device that makes one's colleagues' aircraft symbols fully visible on the radar screen, in addition to those that are one's own responsibility. In this way, Nils has discovered conflicts and has rescued colleagues who "have been into something"; he has alerted his colleagues by saying, for example, "Are you aware of these [two aircraft in conflict]?" Another controller, Yvonne, maintains that most times when incidents have occurred, it has been someone other than the controller responsible who has discovered the conflict.

These examples show how trust is established and used in interactions between colleagues. Such mutual support is necessary so that the controllers can endure the pressure that accompanies their responsibility for avoiding errors or lapses with potential disastrous consequences. There is a collective approach to avoiding failures whereby individual errors may be detected by colleagues and corrected in time to avoid accidents. Behind this achievement lie negotiations, the use of technical means for monitoring colleagues' work, an overlap in competence and a sense of responsibility for each other's work.

Trust and cooperation when new technology is introduced

The probability of dying in an aircraft accident is very low – much less than in automobile traffic and less than within the train traffic system. But air traffic is increasing continuously[5] and the air traffic control system must therefore evolve in order to continue to be safe and also to decrease delays. In addition, the demands for less environmental damage through pollution, noise, etc. are becoming more prominent. From a short-term perspective these demands can be met by, for example, charging higher landing fees in rush hours or at high-intensity airports in order to steer traffic away from these airports.

In a more long-term perspective, however, more radical measures must be applied. Most proposals deal with increasing the controllers' capacity through the introduction of new technology, such as computerized decision support or computer automation of controller tasks. While new technology certainly is needed to deal with increasing traffic, there is a need to analyze potential consequences of proposed technologies. As we have seen, the creation of trust is central to achieving safe air traffic. How will the current processes of creating trust be affected by new technology? I foresee three possible consequences.

First, many of the proposed new systems do not take into account the experience, knowledge and successful working methods that controllers apply today. I will give one example. In an experiment, a computerized decision support was introduced and the radio talk between controllers and flight crews was replaced by an exchange of text messages sent by data link. The system was designed so that the controllers would work separately, one after the other. The controllers therefore lost the opportunity to reach a common solution, using each other's expertise, before they communicated with the flight crews. They also lost the ability to correct each other's mistakes. As we have seen, such procedures are central to creating the necessary trust within the team. These conditions were now lost. Therefore, radio talk should not be replaced by data link, and the use of data link must be carefully designed for those specific situations where it is advantageous and really needed.

A second possible consequence of introducing new technology is the risk that controllers will lose control over what is happening when automated systems take over parts of their work. On many aircraft, technology in the cockpit has taken over a great deal of the flight crew's work. The crew members have become passive managers of automated flight deck systems instead of active pilots (Gras *et al.* 1994; Hollnagel *et al.* 1994). This means that events occur that the crew may not understand. When routinized work is automated, the crew loses the ability to understand and handle more complicated events; they are even less able to handle them when the system breaks down. This is not only a possibility: several accidents have actually happened on such aircraft. The conditions for trust based on understanding and predictability are lost. However, it seems that such problems encountered on flight decks in the early 1990s have been dealt with seriously in newer designs.

Finally, some of the new technologies are designed to intentionally transfer more responsibility for achieving a safe traffic flow to the flight crews. This new division of labor may lead to problems in achieving the necessary understanding and trust between controllers and pilots (Sanne 2001b). Simulation trials of proposed designs for such applications show how controllers in some situations lose the ability to predict and achieve preferred traffic flows. Also, the new technology brings an increased need for mutual understanding between the two communities of practice. To begin with, new phrases must be created and connected to new working modes for both controllers and flight crews, and new ways of interaction between them has to be established, including an increased understanding of each other's conditions in order for trust to be created.

These examples of the potential influence of new technology show the importance of taking into account the controllers' and the flight crews' knowledge and the nature of work before changing the system. The division of labor and responsibility that exists today rests upon negotiations of trust, which are necessary to the achievement of safety and efficiency. These negotiations have to be taken into account in the design of new technology.

Conclusion: creating trust and achieving safety

In this chapter, I have analyzed how safety in air traffic control is achieved through the creation of trust among controllers and between controllers and flight crews. Hazards in air traffic control are caused by the complexity and tight couplings of the system, in terms of simultaneous specialization of tasks and interdependence between different professions and between teams that are often physically separated. Risks produced through complexity are overcome through cooperation and coordination. Tight technical couplings are overcome through a matching redundancy or tight social coupling, i.e. through creating trust. Trust makes it possible for controllers and flight crews to cooperate safely and efficiently and to handle risks that arise as part of their work. It is through communicative and other everyday practices involving an informed use of current technologies that controllers and flight crews demonstrate and acknowledge a common orientation and mutual support, features that constitute the grounds for trust.

Trust here refers to trust in the proficiency of the other. This means that controllers and the flight crews have good reasons to believe that the party they are interacting with is "doing the right thing in this specific moment." The creation of trust involves showing accountability, proficiency and solidarity as well as commitment to others' interests. To this end, the creation of trust is dependent on specific resources at hand in air traffic control, such as current technologies, a distinct organizational culture and specific forms of collaborative and communicative practices.

Trust is displayed and acknowledged through interaction in everyday work, as an integral part of that work. Both display and acknowledgment of trust take place through the everyday use of the technology at hand, such as radar, radio, etc. This means that demonstration of trust is structured by the characteristics of these technologies and that it is only through an informed use of those technologies that trust can be displayed and acknowledged. Thus, technology provides the essential resources for creating trust, but it is not just any use of technologies that generates trust. It is *informed* use, which leaves a trace or a pattern that can be interpreted by others as a sign of the user's proficiency and trustworthiness that is important. To be able to interpret radio talk or radar images and to see such traces or patterns requires competence and experience. Moreover, the trust that is displayed and acknowledged in this way also helps to produce and reproduce the social bonds within the system. These bonds are in turn a resource in the cooperation among the different communities that make up the system.

Finally, while the proposals to design new technologies are necessary to address increased traffic and other demands, it is necessary to consider the means and processes in creating trust and the necessity for trust to handle the risks of air traffic. Unfortunately, some proposals have actually destroyed these means and processes. New technologies that are introduced into the system must instead be designed either to maintain the means and processes for creating trust exampled here – or to create new means to create trust. Knowledge of the complicated negotiations and competences involved is salient to the creation of safety in air traffic, today and in the future.

Notes

1   If not otherwise indicated, all the empirical examples are drawn from Sanne (1999). Some of them have been slightly changed to fit into this text and to make my points more clear.

2   I have also defined the concept in Sanne (1999), based on Heath and Luff's (1992) concept of "interrelated attention" and Suchman's (1993) discussion on the demands on the teams working in "centers of coordination." Common orientation is related to but more fully elaborated than the concept of situational awareness, which is primarily a cognitivistic concept that is related to what the operator knows or should know about a specific situation.

3   It should be noted, though, that although this phrase is not part of the phraseology, neither is it merely a courtesy, since it is recommended by the ICAO (International Civil Aviation Organization) that controllers should try to explain such features of their work to make sure no mistakes are made and to ensure cooperation.

4   This story was forwarded to me by Christer Eldh from one of his fellow ethnologists (thanks!). It should be noted that such stories are to be compared with urban myths: they are not necessarily true, but they nevertheless convey a cultural meaning. For more on this see Sanne (1999: 251 pp.).

5   Traffic volumes have increased most years after the World War II: air traffic has decreased as a consequence of the terrorist attack on September 11, 2001, but it might recover (this was written in March 2002).

References

Bergman, P. (1995) *Moderna lagarbeten: Arbete, teknik och organisation i högteknologisk processindustri {Modern Teamworks: Work, Technology and Organization in High Technology Process Industry}*, Lund: Arkiv.

Drew, A. and Sorjonen, M.-L. (1997) "Institutional dialogue," in T.A. van Dijk (ed.) *Discourse as Social Interaction*, London: Sage Publications.

Eurocontrol (1996) *Integrating FMS Data into Air Traffic Control. A Prototyping Exercise Exploring the Operational Consequences and Possible Effects on Controller Roles*, EEC Note no. 28/95, EEC Task AS09.

Gras, A., Moricot, C., Poirot-Delpech, S.L. and Scardigli, V. (1994) *Faced with Automation: The Pilot, the Controller and the Engineer*, Paris: Publications de la Sorbonne.

Harper, R.H.R. and Hughes, J.A. (1993) "What a F-ing System! Send 'em All to the Same Place and then Expect us to Stop 'em hitting," in G. Button (ed.) *Technology in Working Order*, London: Routledge.

Heath, C. and Luff, P. (1992) "Collaboration and Control: Crisis Management and Multimedia Technology in London Underground Line Control Rooms," *Computer Supported Cooperative Work* 1: 69–94.

Hirschhorn, L. (1984) *Beyond Mechanization. Work and Technology in a Postindustrial Age*, Cambridge, MA: MIT Press.

Hollnagel, E., Cacciabue, P.C. and Bagnara, S. (1994) "Workshop Report: The Limits of Automation in Air Traffic Control and Aviation," *International Journal of Human–Computer Studies* 40: 561–566.

Hughes, T.P. (1986) "The Seamless Web – Technology, Science, etcetera, etcetera," *Social Studies of Science* 16(2): 281–292.

Hutchins, E. (1995) *Cognition in the Wild*, Cambridge, MA: MIT Press.

Morrow, D., Rodvold, M. and Lee, A. (1994) "Nonroutine Transactions in Controller–Pilot Communication," *Discourse Processes* 17: 235–258.

Perrow, C. (1984) *Normal Accidents: Living With High Risk Technologies*, New York: Basic Books.

Randall, D. (1994) "Teamwork and Skill," unpublished paper, Manchester, UK: Manchester Metropolitan University.

Sanne, J.M. (1999) *Creating Safety in Air Traffic Control*, Lund: Arkiv.

—— (2001a) "Interpreting Radar – Seeing and Organizing in Air Traffic Control," in H. Glimell and O. Juhlin (eds) *The Social Production of Technology: On the Everyday Life with Things*, Göteborg: Göteborg Studies in Economics and Business Administration.

—— (2001b) "Simulation as Learning, Enrollment and Organizational Politics," paper presented at the Conference of the Society for the Social Study of Science (4S), Boston, Massachusetts, 1–4 November 2001.

Suchman, L. (1993) "Technologies of Accountability: Of Lizards and Aeroplanes," in G. Button (ed.) *Technology in Working Order*, London: Routledge.

Weick, K.E. and Roberts, K.H. (1993) "Collective Minds in Organizations: Heedful Interrelation on Flight Decks," *Administrative Science Quarterly* 38: 357–381.

8   The social construction of safety in the face of fear and distrust

The case of cardiac intensive care

Sabrina Thelander

How is safety established and maintained in a high-tech environment? This chapter deals with the establishment of safety in cardiac intensive care units, where errors or medical complications can have immediate and disastrous consequences. In order to provide safe care, nursing staff must be able to both predict and avoid risks and hazards. Their job is to help patients who have recently undergone major surgery return to lives that resemble those they lived before they got sick.

Half the people who die in Sweden die of cardiovascular disease, which is the most common cause of death in the country. Nevertheless, current trends show a decline in mortality associated with those diseases; specifically, the mortality rate declined by more than 20 percent between 1987 and 1997. Interestingly, the decline in mortality has been greater than the decline in new outbreaks of the disease, which means that medical care provided to people affected by these diseases has in all likelihood contributed to the improved survival rates. From such a perspective, cardiac intensive care may be regarded as a successful endeavor, even though professionals in this field must deal with greater risks and uncertainty than in many other fields.

Cardiac intensive care involves a high degree of unpredictability, complexity and uncertainty. Some kinds of uncertainty are more intense than others, and I will mention four of them here. First, there is always a *risk that the patient's condition will change rapidly*, which in the worst case may lead to death. Patients in cardiac intensive care units (ICUs) are very frail and nursing staff – despite their expertise, experience and technical aids – can never have complete control over or predict everything happening in the patient's body.

Second, nursing staff in cardiac intensive care units must not only care for critically ill and injured patients, they must also *handle a variety of sophisticated technical artifacts*. Technology is so important in this context that intensive care and sophisticated medical devices are often equated with each other, which can be misleading. As sociologist Robert Zussman (1992) points out, the ICU is first and foremost a place where there are many people and machines. It is also a place where the devices are critical to human ability to manage any medical problems that may arise. Ventilators, oscillators, ECG machines, cardiographs, infusion pumps and blood circulation pumps are just a few of the artifacts used. These machines impose great demands with regard to medical–technical expertise on the part of those who use the devices to monitor and treat patients.

A third form of uncertainty stems from the fact that *a great many people are involved in working with the patients*. There are regulations and expectations concerning staff expertise, and areas of responsibility are clearly defined. Thus there is a division of labor and responsibility that requires individuals and groups to coordinate their actions. These individuals have different training, expertise and experience, as well as differences in their ideas about how the work should be performed and who should do what. The differences are an inherent uncertainty in the work. Knowledge and experience must be transferred and shared and medical professionals must determine what can be expected of their co-workers.

Fourth, uncertainties arise because *quick judgments and decisions* are sometimes required in situations where an error in a cardiac intensive care unit may rapidly have serious consequences. One error in judgment and wrong action may lead to severe medical complications that jeopardize the patient's safety (or in the worst case are so severe that the patient dies). Errors may also have legal consequences for those who work in the unit.

Cardiac intensive care is thus a resource-intensive, high-tech activity that not only requires medical expertise and the capacity to provide care, but also social skills and the ability to use a range of technological artifacts. Several individuals and groups must jointly handle both potentially risky technology and critically ill people. Safety in this environment is the possibility and the capacity to make independent, relevant and competent interpretations of signals from devices and patients. To make these interpretations, nursing staff must have some degree of medical expertise and knowledge of what the various numbers, curves and values on the various charts and recording equipment signify. Another prerequisite is knowledge about every device and every patient, as well as knowledge of how disease trajectories and trajectory work usually develop in the unit.[1] Safety is also related to expertise among nursing staff regarding the management of medical devices and their relationships to the bodies of patients (e.g. knowing how to turn a machine on and off, understanding its function and knowing how bodies and machines must be connected and disconnected to and from each other).

Safety is not, however, established solely in the relationship to patients and the sophisticated medical devices that are connected to them, but also in social interaction, that is, in how co-workers relate to one another. How this process unfolds depends in part on existing organizational frameworks, which means that the nursing staff's interactions influence and are influenced by power structures. The process also depends on various situational circumstances that constitute boundaries and restraints upon the actions of nursing staff depending on how they choose to interpret and thus define the situation. Human actions are situated actions and when they become problematic in some respect, socially constructed rules and procedures are formulated and made explicit. This makes the rules and procedures useful as a basis for reflection and consideration (Suchman 1987: 53f.). In their interpersonal relationships, nursing staff re-establish safety by acting in a manner that serves to present the unit as a whole as a safe place and the individuals who work there as adequately competent. The staff establish and maintain an atmosphere of safety.

We will now take a look at how this sequence of events takes place in two different situations: (1) when someone shows a fear of making a mistake in the trajectory work and (2) when someone does not have trust in a co-worker's ability to perform the work properly. I will be presenting parts of the results of an in-depth case study of two cardiac intensive care units in two university hospitals in Sweden (Thelander 2001). I interviewed licensed practical nurses (LPNs), registered nurses (RNs) and physicians. The interviews were taped and transcribed. At one of the hospitals, observations were conducted for almost four months. When I spent a week at the second hospital towards the end of my fieldwork, I was pleased to discover that the data material felt "saturated" (Hallberg 1994). I have integrated the observations and interviews conducted at the second hospital with the rest of the data material and merged the two units into one that I call "the Unit." My analysis of the material is strongly influenced by grounded theory (Corbin and Strauss 1990). Although the study focuses on the narratives and actions of LPNs and RNs, physicians are highly present in the thoughts of nursing staff, especially RNs, and physicians play an important role in staff work.

In the next section, I analyze why individuals who show uncertainty and fear of making a mistake are treated in different ways: they may be brusquely rejected or they may be treated with understanding and acceptance. We will see examples of situations in which these reactions occur. In the following section, I present various means of managing lack of trust in a co-worker. The alternative I focus on is when the person decides to try to exercise influence over the work but does everything possible to conceal this intent. I describe how this can be done and analyze why it is done. The chapter ends with a few conclusions.

Showing worry and fear

There are many dangers and risks lying in wait in the Unit that can have serious consequences for patients and staff. In a risky occupation such as cardiac intensive care, it is very important that everyone behaves in a manner that promotes safety for all. Those who are calm and do not scurry about, shout or hesitate are perceived as safe – and skilled. But the nature of the work nevertheless causes individuals to sometimes fear making a serious error. To provide efficient and safe cardiac intensive care, they must be able to handle this fear. The nursing staff say that they manage this by individually showing co-workers their fear, after which their co-workers help them and give them the support they need. An inherent tension or problem arises in the trajectory work: at the moment someone shows fear, a conflict may arise between that behavior and the demand for calm and safety. Showing fear does not always coincide with the image one presents in the Unit as an individual and as a group. Co-workers handle the problem in various ways; their behavior proves that they can either reject or support the person who shows fear – and the response depends on *who* is showing fear and *how* it is shown.

People with supervisory responsibility and the authority to make decisions that may have a critical impact on an illness trajectory always meet with acceptance. These people are indispensable to the establishment of safety in the sense that uncertainty or wrong decisions on their part are thought to have potentially devastating consequences. When someone does not meet with acceptance and is instead rejected, that person is, with few exceptions, an LPN. However, all LPNs are not treated equally, and RNs also encounter varying degrees of understanding and encouragement. Why is this so?

Showing fear the right way

There is a recurring pattern in situations where someone shows fear of making a mistake and others in the Unit show understanding. The person who shows fear does so by explicitly talking about what she feels but also by legitimizing her fear. She might mention, for instance, that she has not been given adequate training in using a new device, that she has been on maternity leave and is a little rusty, or that she is new to the Unit. This behavior may be interpreted to mean that the "showing fear" situation has been carefully prepared. Sociologist Erving Goffman (1959) says that one way to avoid situations in which someone's "face" is threatened is to plan carefully and optimally prepare for setting the stage for action. By face is meant the positive social value that a person wants to gain through the actions that others may perceive in the interaction with her. One must show self-control, as well as control the impression that one makes upon others. In the Unit, clearly showing that one is afraid and justifying this fear is an expression of self-control and of control over the others' impressions. Through her actions, the individual emphasizes that her fear – despite everything – is an exception that she does not perceive to be normal.

Even if an individual in the Unit has duly prepared herself for the "showing fear" situation, it is nearly impossible to avoid the threat that this fear represents to her own or the group's face. Goffman describes how people act in situations in which someone's face is threatened, despite all evasive maneuvers, in order to re-establish equilibrium between the actions performed and the projected self. He calls this process a sequence in which the actors make various moves. Goffman identifies four common moves: challenge, invitation, acceptance and gratitude. The participants first notice a behavior that threatens someone's face. They then assume responsibility for correcting the incident – everyone takes on what Goffman calls a *challenge*. Someone, often the person who performed the threatening action, is *invited* to set things right, which naturally means that the action must be redefined. For instance, the person may attempt to make it all seem like a joke, an entirely unintentional action, or something done under duress or strong influence from another source. The efforts to redefine the action cannot stop before "the guilty party" has *accepted* the invitation and thus contributed to re-establishing equilibrium. Finally, he or she should, in some way, show *gratitude* for having been forgiven.

Situations in which fear is accepted in the Unit may be described as such a sequence, but the actors' moves are not taken in the order described by Goffman. In these situations, the person who shows fear takes responsibility for much of the initiative in the sequence that follows. She makes sure her co-workers are aware that she is performing an action that may threaten her and the group's face and invites the others to participate in the face-saving effort. The individual does not, however, just make her colleagues aware of the problem, but also immediately offers a redefinition of her own action so that it appears legitimate and comprehensible – it is recast as an aberrant, abnormal action. This is accomplished when the individual explicitly expresses her uncertainty and need for help. Such action has proven to be a condition for the others to accept the invitation and assist in redefining the threatening action. By immediately accepting the blame for an action that threatens the desired perception of the Unit and the people who work there, the individual improves her chances of being forgiven. The last part of the sequence does not appear distinctly in these situations, i.e. that the person who performs a threatening action finally shows gratitude. The gratitude may be indirectly expressed in the humble, beseeching, almost subservient attitude often taken by the person who performed the threatening action and then tries to redefine it.

"Ugh! No!": setting up a device

One afternoon, Kerstin, an LPN, was asked to set up an infusion pump. She had gone through an orientation about the pump but had never done the set-up before. She had to practice because she was supposed to start working on her own the following week. Several others and I gathered around her as she began:

> The training nurse stands alongside Kerstin and gives instructions. Kerstin is crouched next to the pump. She asks constantly what she should do and I understand that the operations that seem rather simple to inexperienced eyes are not at all uncomplicated to perform. She seems anxious and asks for guidance repeatedly. The atmosphere is noticeably tense. She says "ugh!" and "no!" several times. The training nurse answers patiently, over and over again. (February 21, 1996)

This was a learning situation of a formal nature due to the presence of the training nurse. Kerstin did nothing to hide her uncertainty about whether she could manage the device or the feelings of discomfort or nervousness she seemed to be experiencing. There seemed to be implied permission to ask questions and show uncertainty and anxiety, and Kerstin took advantage of this acceptance. Afterwards, I asked Kerstin and one of the LPNs who were standing by, listening to her and the training nurse, what the consequences would be if the pump were not set up correctly. Kerstin said that it wouldn't matter because an alarm would go off. I wondered then what made the situation so emotionally charged. Why did she express her hesitancy and anxiety so strongly? To explore this question, in the following section I will discuss the dynamics of similar situations and the factors that contribute to them.

"I haven't dared!": using the new CEPAP

In another situation, a doctor told Susanne (an RN) to use a new model of a technical device known as a CEPAP. A CEPAP is a machine that creates positive pressure when the patient breathes so that the patient has to work at breathing, thus expanding the lungs and airways. The Unit has two kinds of CEPAP machines, one that has been used for a long time and one that is rather new. Susanne had to use the new one because the old one was unavailable:

> Susanne talks to the nurse at the next bed. "Our old CEPAP is being fixed? I haven't used this new one – I haven't been trained to use it. I don't know what to do with it, so I haven't dared use it! Do you know what to do?" The other nurse nods and says that she is in the same boat and also feels uncertain about it. Susanne says: "I'll go and get the old CEPAP!" She comes back with it in a few minutes and places herself in the far corner of the room still holding the old CEPAP. She stands there for a long while. (June 22, 1998)

In this instance, the nurse emphatically showed her anxiety about making a mistake and was given sympathy. Susanne showed her almost total helplessness concerning the new CEPAP, while in various ways legitimizing this helplessness. The other nurse showed understanding for Susanne's fear and also expressed the same problem herself. Even if Susanne could not get any practical assistance from her, she received acknowledgment that she was not alone in her anxiety and that this anxiety was no cause for shame.

"It feels strange and unfamiliar!": managing a faulty blood transfusion pump

One afternoon, a new patient was admitted to the Unit. Staff discussed the patient when they were assigning patients to the staff during the weekly report in the nurses' station. RN Stina said immediately that she would very much like to have the patient, because she had been on maternity leave for several months and it had been a long time since she had taken care of that type of patient. The work with the patient later proved very intense. At one point, two LPNs, two RNs and two doctors were gathered around the bed, all of them busy. The patient had been assigned to Stina, which meant she was responsible for supervising the other staff, a role she seemed to have difficulty assuming. She was very quiet and devoted her energies mainly to the blood transfusion pump, which wasn't working well. She and one of the doctors worked with it for long periods of time, while the other nurses and doctors who discussed what should be done. The patient's condition became steadily worse and after an hour of effort, he was taken back to the operating table. (After the patient was taken back to surgery, staff agreed that the pump was the reason the situation could not be straightened out. Several shook their heads and said that they didn't understand why it was used at all.) At one point, the following situation unfolded:

> Stina looks very stressed. She and the doctor fiddle with the blood transfusion pump, which does not seem to be working well. She doesn't say anything, just looks more and more fatigued as time passes. Suddenly, she expresses her uncertainty and fear of not being able to handle the situation. She says straight out to the doctor that the pump feels strange and unfamiliar, that she has been away for so long. He looks at her and tells her, calmly and slowly, that she should *not* be worried, that they are going to help her and that he is sure she will manage beautifully. (September 8, 1997)

Stina ended up in the midst of a situation with a patient in critical condition, a blood transfusion pump that was not working properly and a supervisory role to which she was not accustomed. Finally, she expressly said that the situation was getting the best of her, mainly because she was out of practice. She looked for help and support from the doctor and she got it. Even though he had his hands full trying to save the patient, he took the time to calm her down and help her regain her confidence and to show that he trusted her. In this situation, it was important that Stina acted as a decisive nurse and supervisor and the doctor acted to help her do that. His response seemed to have the intended effect. After the doctor encouraged and supported her, Stina's manner changed noticeably: she talked more, faced the other staff directly and sometimes made specific requests for what should be done. She became the competent supervisor that she was expected to be.

The behaviors exemplified above contribute to safety in the disease trajectories in the Unit. Feelings of inadequacy and anxiety were expressed explicitly. The individuals showed that they wanted understanding and help from their co-workers and they got such support. That is not, however, always the case.

Showing fear the wrong way

There is a discernible pattern in situations where someone shows fear that something will go wrong but is rejected. The person who shows fear does not act to make the others aware that she is behaving in a way that could be offensive, and then she attempts to justify her fear and insecurity. Instead, she tries to persuade others that there is an "objective" problem that must be handled. As I see it, she does not apologize and then take the necessary measures to correct the offensive action. The person who shows fear loses control over what she expresses in the encounter with others and does not invite them to get involved in *facework* (Goffman 1959), in the effort to defend and save her *face*. That is a punishable offense.

Co-workers may react with explicit irritation, but often they simply turn their backs on the person or leave the room without answering any questions. This may be interpreted to mean that co-workers are avoiding the embarrassment of witnessing someone else's failure to act in a manner consistent with the *face* to which she and the group lay claim. But when they do not accept someone's fear, these co-workers also fail to redefine the offensive action as an acceptable one, that is, an action that is aberrant yet justified. Instead, the action seems only offensive and aberrant and the person who performs it appears ignorant or embarrassing.

How should technical information be recorded?

LPN Karin was going through orientation to the Unit. She gave an impression of being anxious. She spoke to no one unless spoken to and spent long periods working with the monitoring curve, which she seemed to spend most of her time filling out. I thought she seemed worried about something. She said nothing to anyone about feeling hesitant or unsure, but she tried on several occasions to make the others aware of something she considered a problem. She never received a response to her questions but was instead met with silence or averted faces. When Urban, another LPN and Karin's orientation guide, came up to her at the bedside table, I heard fragmented comments about how Karin thought that the work with the monitoring curve was problematic in some way because the patient's blood pressure constantly rose and fell. Urban nodded absent-mindedly and listened but did not seem particularly interested. He finally turned away without having said a word. A while later, the RN asked Anna (an LPN) if all the pressures had been recorded. Shortly thereafter, Anna approached Karin and asked her whether she had changed the pressure on the monitoring curve:

Anna: Has this been changed?
Karin: Mm, yes.
Anna: Try to put a slash so that there is a PA in front of it, so it will be like the curve so we will get . . .

Karin leans forward over the curve, interrupts Anna and says without looking at her: Let's see . . .
Filling out the monitoring curve seems very problematic for Karin due to all the different pressures. She whispers to herself a little, then says to Anna: This is a little strange too . . .
Anna says nothing. She turns away and arranges the patient's blanket. Karin falls silent, holds the pen, and looks down at the monitoring curve. Anna goes over to the next bed. (January 20, 1998)

Karin showed clear signs of uncertainty, but she said nothing about feeling unsure. An alternative approach to managing the situation would have been to point out that she was new, had never done the job before, and really needed Anna's help. When another LPN started her shift later that afternoon, she discovered several peculiarities in the monitoring curve, including things she believed were errors.

Why is the alarm on the injection pump going off?

One afternoon, Birgitta (an LPN) and I were alone in a room for only a short while, but during that time, the alarm on an injection pump that was set up next to one of the beds went off several times. Birgitta called out to the nurses' station for help, but no one was there. She looked at me out of the corner of her eye but turned her back and mumbled when I asked whether there was a problem. Finally she left and went to the employee lounge. Shortly thereafter, Birgitta and a few others came back and the alarm on the pump went off again:

> 12.50 p.m. Birgitta goes up to the pump and shuts off the alarm. "Wow, that went off early!" she says and sounds surprised. RN Sinnevi, who is working with the patient in the next bed, looks over her shoulder towards Birgitta and says: "Well, it's old you know, and it's a [brand name]." Sinnevi seems noticeably uninterested. Birgitta says nothing more.

> 1.05 p.m. The alarm on the pump has gone off a couple more times with no one, other than Birgitta, taking any notice. Another LPN and two RNs are also in the room. When the alarm goes off again, Birgitta asks: "Sinnevi, when was it I said something about the pump?" Sinnevi answers: "Probably about 20 minutes ago," then turns and leaves. The training nurse, who has just come into the room, approaches Birgitta. For the first time, Birgitta gets a response to her attempts to start a discussion about the pump and its alarm. They talk for a bit about what may be causing the alarm to go off, then the training nurse leaves.

> 1.30 p.m. The alarm on the pump has gone off a few times. Birgitta looks at the others out of the corner of her eye but says nothing. When

an RN who had been at lunch comes in, Birgitta mentions the injection pump to her. She points out several times that the pump is acting strange, but the nurse doesn't seem to care. Instead Birgitta finally says, irritated, "But it can't be anything, some of them go off even earlier!" Birgitta seems to give up and does not ask anyone about it again, even though I see her fiddle with the pump on several occasions when the alarm goes off. (November 25, 1996)

Injection pumps are set to sound an alarm when there is a certain amount of a drug left in the syringe in the pump. The alarm warns the person monitoring the patient that the drug is almost gone so that a new syringe can be prepared and quickly replaced. Alarms are set at different levels depending on which drug is being administered. If the drug is to be administered at a rate of 1 ml/hr, the alarm is set to go off when there is very little of the drug left in the syringe. If instead the rate is 30 ml/hr, there is much more left in the syringe when the alarm is triggered. Could it be that in the case just discussed there was supposed to be a lot of the drug left in the pump when the alarm goes off and that Birgitta didn't know it? In any case, it seemed to surprise her that the alarm went off when there was still quite a lot of the drug left.

Birgitta tried several times – in various ways and with different people – to initiate a discussion about the injection pump. The only person to pick up the thread was the training nurse, whose tasks include helping nursing staff with technical matters. Birgitta never expressly said that she felt anxious that there might be something wrong. Instead, she tried in various ways to get the others to notice something she perceived to be a problem. She did so in vain – until finally, one of the nurses clearly showed her irritation.

In summary, in order to behave in a manner considered safe, the nursing staff must be aware of and master the "traffic rules of social interaction" (Goffman 1955: 216). Behaving in a safe manner includes always acting so that calm, safety and decisiveness appear to be the normal state of affairs in the Unit and among those who work there. The reactions of people in the environment depend on *how* the person who shows fear conducts impression management (Goffman 1959). This means that nursing staff act so that their behavior seems consistent with the image of the Unit and themselves that they present to each other and to the people around them (including me). The constructed understanding of the Unit is that it is a safe place for staff and patients. Safety is guaranteed by everyone being in control and knowing what needs to be done next. This perception may be threatened by a situation in which someone who works in the Unit shows signs of fear. To show fear for safety's sake may thus lead to doubts about the individual's or the group's face. When that occurs, the nursing staff have a joint responsibility to perform facework, which means (1) that they avoid creating situations in which an individual or the group risks losing face, and (2) if losing face occurs anyway, attempting to defend and save the individual's or the group's face.

Managing lack of trust

Everyone in the Unit has personal experience of and expectations for how the work in an illness trajectory should be performed. They know that they and their co-workers have strengths and weaknesses and that a necessary part of the trajectory work is to monitor others who are involved in a trajectory and judge their competence. In this context, competence means "an individual's capacity to act in relation to a particular task, situation, or context" (Ellström 1992: 21). Sometimes, the act of monitoring leads people to feel that they cannot rely on someone else's ability to handle some part of the trajectory work in the way they desire or expect. They may have doubts about whether the other person knows how to use a device safely, how to make a diagnosis or how to manage medical complications.

Doubt and uncertainty arise in such situations. What can the individuals do in an environment where uncertainty *must* be handled to avoid the risk of a negative impact on an illness trajectory? The loss of trust must be managed within the frameworks of a hierarchical organization with strong power structures (Fox 1992; Franssén 1997; Hughes 1988; Lindgren 1992; Mackay 1993; Peneff 1992; Wagner 1993). To express doubt about the capacity of a co-worker with a superior position may of course be difficult.

The power structures in the Unit affect how nursing staff manage situations in which they distrust a co-worker's ability. The staff's actions are also influenced by whether they believe that the co-worker's actions may have serious consequences. Depending on the balance of power among the individuals involved, their respective positions and the patient's condition, staff may employ three different strategies for handling loss of trust. They *explicitly express their distrust* under two conditions: either when they believe that safety in the trajectory is seriously threatened or if the person in whom they do not feel complete trust is subordinate in the formal hierarchy. If the other person has an equal position or especially a superior one, those who feel distrust simply *do not express their thoughts*, provided that the problem will not have a palpably negative effect on the trajectory in their view. Should that be the case, they try to exercise influence over the trajectory by *covertly influencing the co-worker's actions and the course of the trajectory*.

Nursing staff can lose trust in a co-worker when in their view the latter does not act in the usual or expected way or seems uncertain but says nothing. What do people do when they want to covertly exercise influence over the trajectory effort and why do they conceal their intentions?

Trying to exercise influence: but hiding it

A common strategy used by subordinates to manage loss of trust is to try to exercise influence – but hide this attempt. This strategy may be interpreted to mean that the individuals perform facework by bowing to the formal hierarchy. In this way, they attempt to avoid situations in which the superior co-worker's activities could appear inconsistent with her role and threaten her face. Goffman (1959, 1961) says that one way to avoid such threats is to avoid certain topics of conversation or activities.

In the situations I focus on here, the subordinate co-worker avoids initiating discussion about how a trajectory should be advanced in the safest fashion and the alternative actions that may exist, and she definitely resists expressing doubts about how the other person is acting. By this means, the individuals remain loyal to one of the groups to which they belong, namely those who work in the Unit and provide safe cardiac intensive care. A prerequisite for safe medical care is that each person has the right role and thus the competence to perform the trajectory work safely. The attempt to covertly influence what is done may, however, also be interpreted as an attempt to exercise power – the subordinate engages in a *power play* (Morgan 1986) to gain influence over what happens without having to take responsibility or risk losing face. In such cases, the balance of power must be handled. Nurses in particular do this by employing a sort of "reversed" power: they can use the way that work and responsibility are organized in the Unit to avoid assuming responsibility themselves and avoid appearing ignorant.

Why hide one's intentions?

One reason that nursing staff do not speak up when they believe that someone is acting in error or inappropriately is that in doing so, they may risk assuming more responsibility than they think they can handle. RN Stefanie touched on the problem when she told me that she is reluctant to question doctors' judgments:

> So then someone new comes in who has never worked in the ICU before and it feels like someone else is controlling his decisions, it can be a little frustrating for us who are doing the work. It feels almost like you are treating the patients yourself, which you're not supposed to do, but then you aren't really ... You can manage it to a certain degree, but then you reach the advanced level and you're left standing there. (Stefanie, May 1, 1995)

Stefanie described how she ended up in situations that were difficult to extract herself from and that seemed frightening. What happens if one suddenly borders on an "advanced level" of competence and there is very little time to waste? And what happens if someone later judges that the patient was treated the wrong way? Stefanie did not want to take the responsibility upon herself and relieve the doctor from taking the burden of responsibility.

A second reason for shying away from influence is that people do not want to do other people's work. Nurses in particular expressed some irritation about being forced to take responsibility in certain situations where they should not have been responsible, and for which they are not paid for taking such responsibility. RN Sylvia discussed situations that can unfold when working with patients and medical devices and the only doctors available are non-specialists on call:

> The doctors that work here every day or very frequently know their stuff, but the temporary on-call docs don't have as much experience. They have almost never been involved in the type of situations that we have here, and we always have to get involved and help them during their on-call hours. And make sure nothing goes wrong. That has been the case several times . . . and it isn't fun. If you look at it in terms of pay, here I am with the responsibility in my lap and earning twelve or thirteen hundred dollars a month and they are getting paid five thousand a month. And there I am, doing their job! That is really no fun. (Sylvia, March 23, 1995)

When the nurses spoke about those situations, they frequently likened them to hostage situations where they were forced to take a level of responsibility they should not and did not want to assume:

> Well, you do it for the patients. But you are really almost doing their (the doctors') jobs. When they stand there and *don't have the guts* to do something. (Siv, June 6, 1997)

With the best interests of the patients in mind, the nurses do not feel they can refuse to help the doctors in this situation.

A third reason for keeping quiet is to avoid the risk of appearing ignorant. It may prove that the situation is something the person did not fully understand – and in the worst case something she should have known about. RN Margareta put it this way when we talked about how she manages situations when she believes a doctor is making a mistake:

S: Well, at some of the places I've worked, you just couldn't question the doctors.

I: What do you do if you think a doctor's action is putting a patient at risk?

S: Well . . . you have to . . . risk questioning the doctor anyway and you feel very vulnerable.

I: So you are taking a risk when you do that?

S: Yes.

I: What kind of risk? What are you risking?

S: Well, it . . . it's . . . well, what am I risking? . . . It varies somewhat in different workplaces . . . You may be risking your own – what can I say? You might embarrass yourself if you have judged things wrong. Or you might harm your relationship with that person, I guess you could say. (Margareta, February 23, 1996)

The excerpt indicates a fourth reason for keeping quiet in the situations I focus on here: that it is a particularly sensitive matter to correct a superior in an environment with a strong hierarchical order and strong expectations for competence. LPN Birgitta thought that she needed to take these dynamics into consideration when she ended up in this type of situation.

> Most of the time it's very open ... open channels of communication, and we work very well with them (the RNs). But sometimes, you know, you have to be a little careful with some of them. If you say something, some of them resent it and get angry. And you don't want that because you might have to work with them the next day. (Birgitta, November 25, 1996)

We have seen that the staff, by hiding their intention in the type of situations focused on here, construct an understanding of the Unit as a safe place to be in.

How do they hide their intentions?

RN Susanne, who used to be an LPN, talked about that period and how difficult it could be if she sensed that an RN was making a mistake:

S: Well, I couldn't put my finger on why it was wrong ... if I couldn't really say to a nurse "this is wrong, you can't do that." But if it was something I felt a little unsure of, I kind of walked on eggs, beat around the bush by saying things like "Oh I see, how did you learn to do that? I'm used to doing it this way," or ...
I: You referred to others who ... yes, I see.
S: So I showed them that "I don't understand this myself, but is this situation instead something else?" (Susanne, June 24, 1998)

The situation described by Susanne can be interpreted in several ways. Susanne acted to gain influence over the trajectory work by trying to influence how the nurse performed her work. She did so by referring to her knowledge of how others usually did it and thus suggested that she had knowledge of and experience with it. At the same time, she avoided expressing open criticism. Susanne said she had to "walk on eggs" or "beat around the bush." In other words, she did not say straight out that she thought the other person was wrong. A pattern can thus be discerned in the narratives of nursing staff: it is important to emphasize that one does not know as much or more than one's superior. I interpret "beating around the bush" to be a means of increasing one's personal influence over the trajectory without having to assume more responsibility, risk being criticized or jeopardize one's relationship with the superior.

RN Malin expressed something similar when we talked about situations in which she believes a doctor is making a mistake:

S: It is very hard for a nurse to tell a doctor that, you know . . . I try to find other ways and put it as an open question: "Is this really such-and-such, and is this right, and what will happen if I do that . . . administer this drug, for instance, and . . . well . . . is this right?" And I try to show test results . . . and point out certain things . . . you kind of do it indirectly . . . "And now it's here and this value is such-and-such and now . . . Is this the right drug?" I can sort of pussyfoot around it, do it indirectly in some way.

I: So you have to be careful not to rub it in that you are the one coming up with a reasonable suggestion?

S: Right! (Malin, June 24, 1998)

Malin also referred to knowledge that would give her the right to question the superior's actions. She showed that she understood the drug, the test results, "certain things" and the patient's blood values and vital signs. Just like Susanne, however, she beat around the bush in an attempt to exercise influence. Instead of saying that she thought the doctor should act in a certain way, she continuously asked the doctor whether each of the various routines or procedures might be done a particular way. By taking the indirect route, she contributed to maintaining the image of the doctor as the person who understands the situation as a whole and makes the critical decisions.

Several nurses commented that they are able more easily to indirectly influence superiors because they are women. RN Sigrid, who had just talked about how everyone is free to speak up if they think someone else is making a mistake, noted:

S: Well, you don't say "Excuse me! I think you're wrong, even though you're my boss!" You might put it a bit softer . . . "You know, I feel a little unsure about whether she is doing well enough or . . . do you think you could stand over here?" You know that feminine wile – you make them feel like the big, strong daddy standing there and *then* maybe . . .

I: Yes, I see!

S: Oh yes, a little mixture one thing and another . . . if you can. Not that you chastise someone, there's no reason for that.

I: You have to be somewhat indirect.

S: Yes. (Sigrid, June 24, 1998)

Doctors were often talked about as men, and when the nurses spoke about the type of situations discussed here, they often mentioned their femininity as an asset in the endeavor to gain influence. RN Solveig expressed it as follows:

You have to play on the idea that women are supposed to be a little dumb and say "Can you do it this way too?" (Solveig, April 4, 1995)

In interactions with male doctors, socially difficult situations can be managed by "acting like a woman," by which is meant acting dumb, helpless, unsure of themselves and submissive.

Another means of concealing an intention to exercise influence is to make it seem as if certain actions were performed accidentally. One example is when RN Sassa "resorted to her elbows," namely in a situation in which something that needed to be done for a patient did not get done because a doctor did not know how to use the necessary device. When this type of situation unfolds, Sassa first uses the same reasoning techniques as Susanne and Malin, but if the doctor still does not understand, she acts:

If he (the doctor) doesn't get it, I just have to swing my elbow and say "Oh, look at that, I hit the right button by mistake!" (Sassa, April 28, 1995)

Sassa takes action, which means that she exposes herself to the risk of being held formally responsible if her action leads to a mistake. In such a case, it would not suffice to blame the situation on clumsy elbows. This particular action is thus something that the subordinate only does in special cases when he or she judges that the situation may become critical for the patient.

As we have seen, when nursing staff want to exercise influence over the trajectory work they must take into account a number of situational and structural circumstances such as good working relationships, prevailing organizational hierarchies and divisions of responsibility and labor. These circumstances make it a complicated and sensitive matter to tell a superior that one suspects that he or she does not have the capacity to act. The subordinate avoids taking the credit (or the responsibility) for knowing what to do and still acts according to the principle that the superior has overall responsibility and the right to take initiatives. This action re-establishes the balance of power and assures a smooth working relationship. Nursing staff may also avoid being accountable for transgressing the formal boundaries of their authority or for making mistakes: they act as if they didn't intend to perform the action. Another interpretation of why people hide their intentions to exercise influence over the work is that in so doing, they reverse existing hierarchies: when they think that they can save someone with higher status from making a mistake, this potential gives them satisfaction and a sense that their own position is strengthened (Berner 1999: 89f.).

Finally, there is one last interpretation of nursing staffs' actions that should be mentioned. A subordinate does not try to exercise influence in the above-mentioned way only because he or she wants to avoid assuming responsibility and avoid taking risks. She also does it because the power hierarchies are so strong and resilient that she does not dare act in a way that could risk her own position and relationships to others. The attempt to exercise influence covertly may be a way to circumvent power in the interests of safety and to make sure that an illness trajectory is advanced in the way one believes is best. Nursing staff can thus be said to be exercising "subversive creativity" (Linhart 1982), that is, engaging in an activity that is designed to increase one's personal scope of action in a strongly controlled and structured environment.

Conclusions

At the beginning of the chapter, I asked how safety is established and maintained in a high-tech environment. We now know more about how this is done in cardiac intensive care units in Sweden. I have analyzed how nursing staff handle risk and uncertainty in two specific situations, namely when people show fear of making a mistake and when people distrust a co-worker's abilities. In the first situation, a superior who shows fear and uncertainty is usually given help, while subordinates in the power hierarchy must be able to present their fear and uncertainty as aberrant and abnormal – yet legitimate – or else risk being rejected and ignored. The dynamics of the second situation, i.e. when someone distrusts a co-worker's ability, are different. When a subordinate does not trust the competence of a superior, yet wants to have an influence over what is done with a patient, the distrust must be hidden. The subordinates "beat around the bush" and take the indirect approach in order to handle the situation.

Establishing safety in this high-tech endeavor is a matter of having the ability to exercise influence over the work and therefore over disease trajectories. Safety is also re-established by acting so that the cardiac intensive care unit as a whole is represented as a safe place and the individuals who work there are represented as competent. We have seen how nursing staff re-establish rules of interaction concerning who should be viewed as the one having the central competence to manage the uncertainty that is inherent in all trajectories. Individuals and groups who enjoy superior positions and status are ascribed with the competence most critical to the progress of the trajectories. Medical competence determines social relationships: the higher the position, the more medical competence one is expected to have and the more irreplaceable one is considered to be. By this means, those involved in the situation both reproduce existing power hierarchies and re-establish a perception of safety and security in the work. This perception is essential: if people stopped presuming that individuals and groups in the Unit possess the necessary competence and are capable of performing the actions expected of them, the Unit would become a dangerous place in which to be a patient and staff.

Note

1 The course of a patient's illness as it progresses over time, as well as the actions and interactions that contribute to the course, may be seen as an *illness trajectory* (Strauss

et al. 1985). This trajectory is shaped by things such as social and economic conditions, as well as by interactions among nursing staff. These interactions have an impact on how behavior is coordinated, how tasks are allocated, and on how individuals position themselves in relation to one another (Strauss 1993). The medical work may be described as a constant, collective struggle to make a patient case a successful case; an illness trajectory must always be kept under control and "moved forward" even as its course is constructed (Berg 1997; Strauss *et al.* 1985). In and through the process, the patient's problem is made into one that nursing staff can manage, that is, a problem they feel they can handle with the help of existing procedures and ideologies (Berg 1997: 127).

References

Berg, M. (1997) *Rationalizing Medical Work: Decision-support Techniques and Medical Practices*, Cambridge: The MIT Press.

Berner, B. (1999) *Perpetuum mobile? Teknikens utmaningar och historiens gång*, Lund: Arkiv förlag.

Corbin, J. and Strauss, A. (1990) *Basics of Qualitative Research: Grounded Theory Procedures and Techniques*, London: Sage.

Ellström, P.-E. (1992) *Kompetens, utbildning och lärande i arbetslivet: Problem, begrepp och teoretiska perspektiv*, Stockholm: Fritzes.

Fox, N.J. (1992) *The Social Meaning of Surgery*, Buckingham: Open University Press.

Franssén, A. (1997) *Omsorg i tanke och handling: En studie av kvinnors arbete i vården*, Lund: Arkiv förlag.

Goffman, E. (1955) "On Facework: An Analysis of Ritual Elements in Social Interaction," *Psychiatry* 18(3): 213–231.

—— (1959) *The Presentation of Self in Everyday Life*, Harmondsworth: Penguin.

—— (1961) *Encounters: Two Studies in the Sociology of Interaction*, Indianapolis: Bobbs-Merill.

Hallberg, L. (1994) *En kvalitativ metod influerad av grounded theory-traditionen* (Psykologiska institutionen), Göteborg: Univ.

Hughes, D. (1988) "When Nurse Knows Best: Some Aspects of Nurse/Doctor Interaction in a Casualty Department," *Sociology of Health and Illness* 10(1): 1–22.

Lindgren, G. (1992) *Doktorer, systrar och flickor*, Stockholm: Carlsson förlag.

Linhart, D. (1982) "Au-dela de la norme," *Culture Technique* 8: 91–98.

Mackay, L. (1993) *Conflicts in Care: Medicine and Nursing*, London: Chapman and Hall.

Morgan, G. (1986) *Images of Organization*, Beverly Hills: Sage.

Peneff, J. (1992) *L'Hôpital en urgence*, Paris: Éditions Métaillé.

Strauss, A. (1993) *Continual Permutations of Action*, New York: Aldine de Gruyters.

Strauss, A. *et al.* (1985) *The Social Organisation of Medical Work*, Chicago: The University of Chicago Press.

Suchman, L. (1987) *Plans and Situated Actions*, Cambridge: Cambridge University Press.

Thelander, S. (2001) *Tillbaka till livet: att skapa säkerhet i hjärtintensivvården*, Lund: Arkiv.

Wagner, I. (1993) "Women's Voice: The Case of Nursing Information Systems," *AI and Society – Special Issue on Gender and Computers in a Cultural Context*: 1–20.

Zussman, R. (1992) *Intensive Care: Medical Ethics and the Medical Profession*, Chicago: The University of Chicago Press.

9 Constructing workplace safety through control and learning

Conflict or compatibility?

Marianne Döös and Tomas Backström

> When risk enters as a concept in political debate, it becomes a menacing thing, like a flood, an earthquake, or a thrown brick. But it is not a thing, it is a way of thinking.
>
> (Douglas 1992: 46)

Introduction

The bulk of safety work in today's working life in Sweden is rooted in a long and successful tradition developed in the course of moving from heavy and risky work to a situation where many physical hazards of the work environment – including accidents – have been eliminated or substantially reduced (Elgstrand 2001). The success of this tradition relies on important standards, rules, investigations and inspections that have been defined and mandated by two main authorities, namely the Swedish Work Environment Authority and the Work Environment Inspectorate.[1] This governing and controlling rule system has been developed in collaboration with trade unions and employers' associations on the national level.

In recent years, Swedish working life has, however, arrived at a situation where demands for *control* on the one hand, and demands for *learning and acting competently* on the other hand, appear to come into conflict – for organizations, their managers and their members (Sitkin *et al.* 1994). Many companies in Sweden and elsewhere act in a market in which the critical competitive factor for success is not only competence but also its development and renewal. Firms must, therefore, try to organize an entire business and its daily activities so as to promote learning and development of competence. Individuals and work teams are increasingly supposed to function with responsibility and autonomy, interpreting internal and external signals in order to change ways of working. These kinds of changes require that safety is increasingly constructed in the day-to-day performance of work tasks, encompassing aspects that cannot be taken care of through targeted safety work alone. Business-related and production demands for *learning* call for thought and reflection: posing one's own questions, pondering and taking initiatives, doing active experimentation, challenging existing problems and forgiving error. The occupational safety of firms is, however, still managed through the principle of control. In contrast to learning, *control* requires and creates a certain kind of culture of obedience, built on a foundation of rule-following, fixed routines, negative feedback loops, regular activities and repetition.

An emerging problem for safety management can thus be described in terms of two incompatible (or at least seriously conflicting) principles – incompatible because of the consequences they have for defining work tasks and people's actions. We refer to two ways of cultural functioning, that is, learning versus control as the basis for safety work, that rely on different and opposing principles. What one of these requires, threatens at least in part the other, and vice versa. If safety is built on control, while the main task of the business in its ongoing production and its quality and improvement is organized along principles of learning, then the ordinary employee is at risk of becoming cognitively and emotionally "caught between the cultural expectations" (Watson 1977: 3). The worker is caught between conflicting demands for quite different abilities, actions and ways of thinking and functioning.

Is such conflict just an academic issue? Maybe not. Based on lengthy experience in the arena of occupational safety, our current sentiment is that *we are now encountering a problem of safety in contemporary workplaces, since the production, quality and improvement work is now turning to being governed by principles other than those for which work safety measures were initially designed and in tune with.*

The indications that we have noted so far and that are related to risk and risk management might mark the beginning of a negative accident trend. Extrapolating a trend where part of the business says "Think and act," whereas safety relies on "Obey and follow rules" is to anticipate a situation in which accidents again occur in larger numbers. Of course, this vision is an extremely crude picture, lacking detail or variety. Our purpose is to offer the reader a simplified description so that differences can be captured and borne in mind in light of the reasoning that follows. This chapter thus aims to address the conflict of principle through describing the conflict between learning and control in safety work, its possible consequences, and the eventual dilemma it presents for actual work practices. Taking an exploratory approach, we will bring together relevant theory and empirical examples in a discussion of discernible trends and in a quest for ways through which control and learning may coexist.

This chapter is organized as follows. We begin by looking at various managerial fields in which similar distinctions have been made between control and learning. Then, in order to explore and discuss the theoretically identified conflict and its practical implications in the domain of occupational accidents, we will then describe principles and ways of functioning in older, traditional, so far successful and reliable, control systems. With the same intent, we present principles and mechanisms of learning among individuals and in an organizational setting. Empirical data from two companies, including information provided by twenty-four safety engineers, are then provided. The concluding section of the chapter discusses the potential for coexistence of what we will refer to as *risk control* and *safety learning*.

Distinguishing control from learning

Researchers in the managerial fields of quality control and organizational development have distinguished between control and learning. Sitkin *et al.* (1994) discuss the example of total quality management (TQM) as a control-oriented approach. TQM is a management strategy which places explicit emphasis on quality. Sitkin *et al.* (with reference to Snell and Dean 1992) state the following: "the core features of TQM as it has come to be prac-ticed: 'total quality is characterized by a few basic principles – doing things right the first time, striving for continuous improvement, and fulfilling customer needs'" (Sitkin *et al.* 1994: 538). TQM is criticized on the grounds that it does not pay attention to the uncertainties that face organizations, and it is described as being "in danger of being 'oversold,' inappropriately implemented, and ineffective" (op. cit.: 538).

TQM has usually been implemented as a form of cybernetic control: it has a feedback loop that measures and compares the output of a process against a certain standard, feeds back information on variances, and amends the system on the basis of this information. Sitkin *et al.* argue that this proce-dure is poorly suited to changeable and uncertain situations. In a changing world, the efficiency of an organization is based on an ability to balance the conflicting goals of stability and reliability on the one hand against ques-tioning and innovation on the other hand. In contrast, control via feedback loops is based on principles of repetition and standardization, implying routine work as a prerequisite, and indicating, as well, the need to know beforehand what different effects a decided action will have. This kind of approach to control is less well suited to situations of great uncertainty, when there is more a need to focus on changing insights or objectives of actions than on changing rules. When innovative thought is more important to competitiveness than reliability, such control approaches are inappropriate.

To clarify the distinction between control and learning, Sitkin *et al.* advo-cate that TQM is divided up into *total quality control* and *total quality learning*, where both have separate objectives that must be focused upon in order for them to coexist. Characteristics of total quality control (TQC) are the use of feedback loops and benchmarking for information on production and customer needs, the enhanced exploitation of existing resources, the use of standards in control and evaluation, and the achievement of zero defects as an attainable target. In contrast, total quality learning (TQL) is character-ized by resilience in the face of the unexpected, innovation and use of new resources, self-designed and changing diagnostic information, and the use of calculated risks. Sitkin *et al.* take the view that a balanced mix of control and learning is required, since "organizational survival in today's highly competitive and rapidly changing world depends both on reliable performance and adaptability" (op. cit.: 540).

Another important area of work concerning control and learning in organizations is recent Swedish research on "continuous improvements" (the Japanese "Kaizen"). Here the organizational development work of companies is described in terms of two kinds of managerial strategies, namely programmatic strategies and learning strategies (Hart *et al.* 1996). Such strategies have basic features that can be viewed as compatible with the distinction between control and learning that is put forth in this chapter. A programmatic strategy includes the a priori setting of solutions by managers, adopting a familiar method for change (rather than an innovative one), viewing change as a time-restricted project that enables leaps to be made to new stable states and believing that prediction is possible. In contrast, the learning strategy requires other fundamental standpoints and "seldom strives to organize operations so that they remain stable" (op. cit.: 121, our translation). It is characterized by recognition of the importance that all members of an organization share a common vision, the presence of participative interpretation and design in which change is seen as ongoing adaptation and improvement, and the belief that communication is an essential activity. In actual practice, as Hart (1999) points out, the choice is not a matter of "either/or" but rather of organizational tendencies – that is, whether a company tends to act programmatically or in a more learning-oriented manner.

Principles of risk control

In organizational strategies that strive to achieve safety through control, instructions and rules are key elements. According to control principles, the anticipated strength of an instruction or a rule lies in its existence and its being followed. Consequently, how a rule is constructed and what its consequences are for an individual or a team are not focused; instead attention is directed to rule obedience and to observable action, which is treated as both predictable and controllable. This approach generates various potential *safety ideals* (Backström and Döös 2000). One safety ideal might be that all hazardous energies (for example, machine movements or harmful substances) which cannot be removed or distantly located should be encased by physical barriers. Another ideal might be for organizations to prepare safety instructions that are so detailed and extensive that, if followed to the letter, accidents would be impossible. Both solutions, however, give rise to a serious conflict between safety and production. For example, production problems that call for corrective action in a hazardous zone may be impossible to handle; the machinery or rules may not be flexible when changes in production are required, and so on. It is well known that production typically emerges victorious from such conflicts. This problem is one of the critical gaps in risk control (Backström 2000).

Further, managing risk through control does not take account of the fact that individuals are intentional in how they define and carry out tasks. Health and safety managers often talk about problems of individual risk taking, such as the tendency that personnel do not follow safety instructions and so on – even in companies with well-functioning risk control. The task of controlling risk is viewed as requiring conscious acts and measures, while normal problem solving within a work task is, to a large extent, not regarded as crucial for the construction of safety. On the contrary, such problem solving tends to be seen as a form of risk taking.

The basic dynamic of risk control is based on the idea of a negative feedback loop. By means of safety rounds, investigations into accidents, risk and safety analyses and similar practices, it is presumed possible to measure the level of safety within the organization. The results of such measurements are then analyzed, providing a basis for formulating action plans and making decisions that will help the work environment to achieve the level of safety that has been stipulated as a goal. This process is usually carried out with the help of some key measure of outcome, such as the accident rate. Standards and routines then offer assurances that the health and safety activities at the workplace are good enough and thus maintain the specified level of safety.

Internal control is another well-described and empirically tested version of risk control. Since 1992 all Swedish companies have been required to employ a specific model of internal control in their health and safety management activities. The ordinance behind this model, entitled "Internal Control of the Working Environment" (NBOSH 1996) requires companies to conduct various kinds of studies or measurements of the work environment, as well as carry out an annual follow-up of internal control. Obligatory routines have to be formulated, and documents describing different hazardous situations prepared; goals with regard to levels of safety for the work environment must be formulated and action plans are mandatory. Kjellén (2000: 15) states that the "internal control regulations in Norway and Sweden require companies to develop and implement adequate management systems to ensure that their activities are planned, organized and maintained in accordance with the SHE legislation" (author's note: SHE legislation is the national law – including ordinances, regulations and court praxis – in the area of safety, health and environment).

Empirical example 1: a deviant body of thought within an internal control approach

In a pilot study to evaluate the above-mentioned Swedish internal control (IC) ordinance, a series of case studies were conducted that aimed at describing IC at a number of companies that according to Sweden's Labor Inspectorate had well-functioning such systems. One of these case studies was performed at a newspaper printing plant (Backström 1998). The printing plant had moved to new premises and modernized its production technology in 1992/1993, and management planned to renew the entire plant operations. Toward this end, an organizational change was effected, and an attempt was made to improve the methods, orderliness and coordination of health and safety management, recognizing that these activities would otherwise be incompatible with the new production technology. Notably, a fatal accident occurred early in this period, vividly demonstrating the significance and urgency of internal control.

The success that the printing plant achieved in strengthening its mechanisms for IC was due at least in part to the involvement of certain "entrepreneurial actors" who promoted these developments. One new manager became head of a department that had not existed previously, which meant that the new department had no ingrained culture or collective structures of meaning (Dixon 1994) that needed to be dealt with. In 1992 the manager was assigned the task of designing an IC system for the company that was simple and readily understood. By 1994, he had prepared two thick files describing the new IC system at the plant. The documents aimed at a more systematic health and safety management. They stated that certain activities such as safety rounds and hazard controls should be performed at pre-specified time intervals, and they provided checklists to guide the IC activities. All problems were to be reported on specially designed forms. Documentation was crucial, and action plans were to be prepared and implemented on an ongoing basis. In addition, all middle-level managers had to undergo a training course concerning the new system. When the Labor Inspectorate performed its first inspection of the plant in February 1997, it assessed the company as having a well-functioning system of internal control (particularly after the company adhered to the Inspectorate's directives for change as prescribed in the course of the inspections).

In some of the printing plant's other departments, however, safety work was *not* pursued in accordance with IC principles. Instead of an emphasis on formal rules and routines, an alternative mode of working had been developed. Here, the ideal of safety work was described as informal and dialogue-based. In the middle managers' view, IC work should be based on managers' having good contact and active dialogues with all subordinates. During a daily talk with the manager, each employee would have responsibility for reporting hazards and problems, and each middle manager would be responsible for immediately taking measures to deal with any identified problems. According to these managers, formal activities and functions (such as documented action plans, safety rounds and the activities of safety representatives) would become superfluous, since virtually everything could be resolved in talks between managers and workers during day-to-day production. Unfortunately, the case study of the newspaper company provided no data to enable an assessment of how well all this has functioned in actual practice. Nevertheless, these middle managers' descriptions of a radically different approach to issues related to the work environment suggest an alternative to IC as a means of attaining safety at the workplace.

These different approaches to IC in a company described as successful with regard to IC illustrate the fact that different departments within the same company may apply contrasting principles and methods in their safety work. This finding suggests that major actors within the company were advocating alternative approaches and that company management viewed organizational learning as something to strive for in its production work. Did managers in the deviant departments perhaps also apply learning concepts in their other operations that related to improving the work environment? A learning-based alternative to risk control is elaborated upon in the following section and referred to as *safety learning*.

Experiential learning and organizational learning: a theoretical perspective

In the many situational work tasks of everyday life, ways of thinking and understanding are continuously constructed. To change one's way of thinking or understanding means to learn. Learning can, in brief, be described as a situated process of constructing knowledge based on action, where the learner is an active constructor of knowledge and know-how (see, for example, Löfberg 1989; Piaget 1970). To learn implies changing one's ways of thinking and/or acting in relation to the task one intends to perform. The outcome of learning can be envisaged as consisting of two aspects. Within the individual, learning is expressed as constructing and reconstructing one's *cognitive structures or thought networks* (Döös 1997). In terms of outwardly visible signs of learning, outcomes are expressed in the form of *changed ways of acting, performing tasks and talking*.

For individuals in an organization, the work tasks, problems and situations of everyday practice afford ongoing opportunities for learning (see, for example, Döös 1997; Löfberg 1996). Individual experiential learning (Kolb 1984) can be understood as an ongoing interchange between action and reflection, where past experiences provide the basis for future ones. Action forges the link between the human being and the environment, where active participation and personal action are prerequisites for the learning process to take place. Through action we enter into both concrete experiences and the reflection that build up our know-how over time; through action we, as people, take part in and change circumstances, conditions and situations.

Within all organizations and work teams individuals learn and then extend their knowledge to the next situation, to other specific environments and meaning contexts, and to interactions with other individuals. We learn together, but this learning is always grounded in understandings that each of us carries with us. In a study of day care and after school centers for children, Ohlsson (1996) refers to a continuously ongoing dialogue in which the personnel adopted "a narrative form, whereby they made their experiences available to each other; joint reflection, when they reflected on each other's experiences and joint intention, where they developed joint strategies to treat a specific child or to cope with more general tasks" (op. cit.: Abstract). Ohlsson's line of reasoning has a strong affiliation with Dixon's (1994) usage of the concepts of private, accessible and collective meaning structures. These concepts refer to different context-specific degrees of being willing and able to make one's own thinking – indeed one's own way of understanding – available to others.

Learning as a collective process thus means that individuals learn through some kind of interactive and communicative action. This is a learning process that creates synergetic effects, whereby what is learned within the work team becomes qualitatively different from what any one of the participators could have reached alone. Furthermore, this conceptualization refers to learning that results in shared knowledge, in similar understanding of something specific, and – grounded in this shared knowledge and understanding – in the individuals' ability to act. In making theoretical sense of collective learning, several theorists have emphasized the importance of dialogue, talk and reflection. Transformative (Mezirow 1990) collective learning takes place, according to Wilhelmson (1998), when group participants communicate their perspectives as well as broaden, shift and transcend them. This process thus entails transcending the individual perspective in that the group forms a new and (at least for the moment) common understanding. In Piaget's (1970) terms, the process can be understood as a kind of collective accommodation.

A review of findings from mainstream contemporary empirical research into safety culture reveals that the *construction* of risk and safety is largely overlooked. Instead, a positivist, reductionist perspective is typically adopted (see, for example, Cooper 1998; Coyle *et al.* 1995; DeJoy 1995; Williamson *et al.* 1997; Zohar 1980). Such an approach entails attempting to break down the complicated concept of safety culture into a number of sub-areas, after which the researcher attempts to identify a small number of variables that can be used as the basis for (unproblematized) measuring of the level of safety culture in an organization. In order to be able to link learning to safety, however, we need to investigate how risk and safety are actually constructed in everyday work situations.

Empirical example 2: constructing risk and safety

Our second case study focuses on an engineering company with several thousand employees. Its physical premises include a foundry and workshops for machine production and assembly. Production takes place in long series and is highly automated. Over the last ten years, there has been a trend toward group-based organization, in which increasing numbers of tasks have been delegated to autonomous work groups. The level of workplace safety is high. There is an in-house Occupational Health Services (OHS) unit that includes a technical section with several safety engineers. The number of accidents has fallen steadily each year and the total accident rate has demonstratively declined in the last decade. At least part of the reason for this decline stems from the extensive safety measures that have been adopted within the company. These measures include the forming of groups to improve the work environment, targeted safety rounds, focused investigations of accidents, risk and safety analyses, employee training, introductory programs for new employees, installments of new and safer equipment, certain measures for preventive maintenance and risk assessments in connection with acquisitions of new machines.

The relevance of theoretical perspectives on experiential learning is evident in a situation such as this one, in which safety cannot rely on management through control because workplace participation has assumed a more direct and autonomous character. Risk (and safety) is constructed when people are not at all consciously dealing with it. Viewed from this perspective, *a situation of high workplace safety does not primarily stem from the implementation specific safety measures, but rather from the way people solve ordinary work tasks and their related problems* (Döös 1997).

From the perspective of constructing safety, one important aspect of an organization's handling of disturbances is whether or not it is prepared to live with known faults. In a number of studies (for example, Backström and Döös 1995), the handling of disturbances to production has been shown to be the task during which the bulk of accidents occur in automated production. In the light of this clear risk situation, Döös (1997) focused on operators' learning in connection with disturbances in a workshop at the engineering company. The findings of this study confirmed first, that people in the workshop did put up with various known faults in the form of difficulties, disturbances and problems for a long time. Second, the study showed that operators with ongoing experiences in this respect, and who thus developed a multidimensional, reflective problem-solving stance, still did not fail to prevent risks. The reasons for this were not, however, related to risk or safety but rather out of concern for production, where the guiding criteria were product quality and quantity – in short, keeping the machines running effectively and according to plan.

In the following, we will provide some commonplace examples of how people think at work – that is, in the context of performing tasks critical to safety – in situations where their concentration and intentions are focused on performing an immediate work task rather than on safety, internal control, or regulations. These examples are based on extensive fieldwork in the engineering company as indicated earlier.

Example 1 – Why was stop not activated? Operators may choose to not use stop functions because they may prefer to let the machine continue so they can observe how it is actually operating. The operators assert that it is not possible to trace a fault with the machine at a standstill, the stop functions available do not permit this type of observation and dynamic testing. Tracing faults and handling disturbances are not purely cognitive tasks. They are performed as an ongoing combination of thinking and manual work, after which one sees the result of what has been adjusted, touched, or blown clean. It can be described as an ongoing dialogue between man and machine.

Example 2 – Why do machine movements come by surprise? When all attention is solely focused on effecting a particular adjustment, it is not surprising that people forget to take care and are easily caught unawares by an unanticipated machine movement. Even if the movement comes at normal speed, it might be perceived as faster or more powerful than normal due to the operator's concentration, it is perceived or noticed in a different way. The focus of the operator's attention is not on what will happen when the machine starts but rather on actually getting it to start. It is the latter problem that is the immediate priority, and in this situation, it is unlikely that the operator will direct his thinking to the risk of e.g. stored energy in machine movements and how best to minimize these risks.

Example 3 – Why was it believed to be a safe area? The study highlights both how one "talks" with the machine as well as how one gets almost personal with it. Despite all safety devices, talking and getting almost personal with a machine can prompt an operator to think that he can rely on both the machine and himself: when you have many daily conversations with the machine and it always replies as expected, some kind of relationship of trust arises. The operator knows what to expect from the machine; it is usually an honest partner that provides clear feedback on actions taken. You usually know! So why stand in constant readiness for the unexpected?

(Döös 1997: 210, our translation)

Operators can thus sometimes more or less rely on a "let's fix it for the moment" approach, without having any intention to permanently solve the problem. In other words, *the risk in question was continually reconstructed – recurrently and non-reflexively – in ongoing work practice.* But this type of temporary patchwork approach also reflects the operator's desire to clear difficulties out of the way so as to facilitate performing one's own work.

One of the factors that shapes how an operator at the engineering company chooses to look at a particular situation or fault is the nature of the fault itself: a fault that constantly recurs, and also causes difficulties for the operator, will be high on the operator's agenda to correct because it is the type of fault that he wants to get rid of. On the other hand, an operator can live for a long time with a fault that seldom recurs and/or is easy to handle once it arises. Also, operators will tend to put up with faults that for various reasons are hard to trace. Here cost also comes into the assessment; fault tracing is expensive and imposes a burden on a production department.

Information about maintenance costs is provided by management at every monthly meeting of the workshop. Also, as time passes, new variations of the fault might turn up, or the operator may notice something he has not seen before – thereby clarifying the situation and allowing the fault to be dealt with on a more permanent basis.

Safety culture and safety learning

We are now ready to attempt a definition of "safety learning." Before pursuing this definition, however, we want to discuss the concept of safety culture, which is a frequently employed concept in the current context.

Over time, an organization's thought networks and ways of working create a specific work culture. Alvesson (1990: 13), referring to Ortner (1984), says that culture should be regarded as "a collectivity's common conceptions, meanings and symbols" (our translation). Culture is maintained and recreated through socialization – that is, through the fine-grained webs of contents and meanings that permeate social practices, as well as through words, activities and phenomena that have strong symbolic value. To a substantial extent, culture is also a matter of shared conceptions of reality which are not given once and for all, but which instead are continuously and jointly constructed and reconstructed in various work tasks and situations within an organization.

From this perspective, culture emerges as a collectively constructed social order, as a collective context of social meaning. As Döös and Ohlsson (1999) have pointed out, contexts of meaning are formed through the activities by which members of an organization shape and assign meaning to their shared and collective practices. *Achieving a healthy safety culture is a process of collective learning and, accordingly, one that requires constant construction and reconstruction at different levels within an organization.*

Both in research and actual work practices, we observe an increased interest in focusing on the question of how a good safety culture is created and maintained. Reason (1997) recently made a number of conceptual distinctions by identifying various aspects of safety culture. One such aspect stems from a distinction between national and organizational cultures: "Whereas national cultures arise out of shared values, organizational cultures are shaped mainly by shared practices" (op. cit.: 192). This appears to be a worthwhile distinction – but only if it is recognized that even a national culture, both originally and on an ongoing basis, is shaped by shared practice (cf. Berger and Luckmann 1966).

For our purposes, the important part of Reason's observation lies in the second half of this quotation. If management is to be exerted with the help of learning, it should be self-evident that the processes by which culture is created and constructed should be borne in mind when organizing everyday work. Lindheim and Lien's (1997) case study illustrates this crucial insight in a workplace in which emphasis was placed on designing and developing a structure for meetings which ensured that participants' different roles and work functions encountered one another.

In efforts to prevent occupational accidents, it quickly becomes evident that talk simply is not enough; yet, in the literature that deals with processes of collective learning, talking seems to be somewhat overemphasized (Döös *et al.* 2001). From a safety perspective, it is necessary to emphasize the link between thought and action, as well as the importance of the kind of learning that emanates from seeing one another's actions and from jointly carrying out tasks. Such a perspective clearly helps to underscore the importance of sense making in action (and sense making as an outcome of action). On the whole, experiential learning theory provides useful tools for understanding the processes by which safety as a real-life phenomenon is constructed, changed and developed. Experiential learning implies a process perspective, while the concept of culture implies a more stable, ongoing state.

By *safety learning*, we mean learning that focuses on the members of an organization as striving towards achieving a safe and healthy work environment, reflecting a collective awareness on their part of the importance of such a work environment, as well as a compelling and continually evolving collective vision of it (cf. the lines of argument in Dwyer 1991). Edelman's (1997) dissertation, for example, shows how safety among railroad depot workers was created through their working together. Safety learning also focuses on an organizational learning cycle in relation to the work environment: that is, most people participate in introducing work environment issues into collective discussion, which in turn generates shared interpretations of problems. The concept of safety learning also means that people are given the authority to act in line with such shared interpretations (Dixon 1994). Accordingly, safety learning can be understood as implying that safety cannot be fully planned but rather that it emerges through interactions between members of the organization, as well as between these members and others in the external environment. A crucial guiding theme for safety learning might be peer concern for their colleagues (Dwyer 1991; Edelman 1997).

Safety and the coexistence of control and learning

In addressing control and learning, we are conceptually situated at the abrasive interface of two distinct cultural principles. Control in itself makes no claim to impact on culture, although perhaps it can be regarded as making such a claim without management's reflected awareness. Control does not imply any thinking in cultural terms and has different preconditions than those for actively constructing safety in the everyday tasks. Like all other activities, control affects culture when it is exerted, but it focuses on the system's relation to the environment rather than the internal properties of the system as culture. Control is implemented within a culture, and control work is involved in creating a culture. In contrast, learning has a direct focus on culture. In our view, the outcome of learning is not something unequivocally structural but rather part of a process.

Although control is both a form of governing and a form of culture, when control is used as a management principle, there is typically no focus on culture at all. There is no thinking in cultural terms, there is an indifference to the processes by which culture is created, and there is no reflection over what one is actually creating: the link between the goal in implementing a particular measure and its outcome in actual practice is not well explored. And when you do not know what you are doing, there is a risk of unintended consequences which are quite different from the anticipated outcome – or consequences that actually counteract the desired outcome. Control may be able to provide good safety, but it collides simultaneously with the active development of skills and independent thinking that many personnel strive for (although perhaps for a variety of other reasons such as quality, customer benefit/entertainment, and so on). Control may be said to focus on doing and behaving, while learning is directed at thinking and acting.

Our view, indeed the very foundation of our approach, is that both managerial principles – those of control and learning – are always needed, although the optimal distribution between them varies with the specific context. In companies that have (1) a traditional, authoritarian style of management, (2) one specific department or division for quality and (3) a tendency to run projects for change over which employees can exert little influence, it is likely that a greater degree of control will be involved in efforts to attain safety. In contrast, companies that have autonomous or semi-autonomous groups which take responsibility for their own work and pursue continuous improvements can be expected to be characterized by a greater degree of safety learning. The choice of approach might also depend on the degree of stability in the company's environment. Companies with a stable market that produces standard products year after year by using essentially the same techniques may be assumed to achieve a high level of safety by means of risk control. In contrast, those that are constantly changing their products and technologies must have extensive safety learning.

Finally, the nature of the risks also influences the extent to which the two approaches are adopted. Hypothetically, we can envisage that so-called "imported" hazards – that is, those hazards that originate outside the organization – are best handled by means of control. For example, such hazards may be handled by strict specifications, inspections of all installations and risk/safety analyses when purchasing equipment. It may also be reasonable to apply a greater degree of control when an operation is on the verge of triggering a risk with catastrophic consequences. On the other hand, everyday risks at the man–machine level (or risks in connection with routine preparations and purchases) may be handled with a greater degree of safety learning.

Empirical example 3: combining control with elements of learning?

In the Occupational Health Services unit of the engineering company previously described, the control principle has been clearly evident. In the spirit of prescriptive regulatory control, the company has had many control-oriented tasks and procedures on a regular basis, such as targeted safety rounds, in-house safety regulations, work injury investigations, work environment reviews and risk surveys. All activities for improving the work environment are formalized and documented. They are based on the competence of experienced and knowledgeable safety engineers; the expert role has been close at hand and has fitted in well with the traditional engineering workshop culture with its hierarchical and bureaucratic structures. Emphasis is placed on rules and routines, and there are clear procedural requirements for many work and safety issues, such as procedures for accident investigations and the handling of information obtained from these. The company has also put considerable effort into making new machines and groups of machines safe from the very beginning. Safety requirements have been included in the specifications given to dealers and suppliers, and these specifications have been followed up by scrupulous inspection when the machines are installed.

Many safety engineers within the in-house Occupational Health Services express the general view that the difficulties with this type of internal control do not stem from the extensive procedures for documentation, but rather from getting internal control actually used. In striving to achieve such control at the engineering company, the safety engineers did not want to intervene solely as experts, neither did they want to provide production departments with finished investigations or ready-made proposals for action. Rather they saw their key task as one of supporting the production departments' own safety work. These safety engineers thus sought work methods and tools that might initiate processes in the various production departments, stimulating personnel and management to jointly identify the risks they faced and articulate how these could be dealt with. This type of initiative would then be provided with access to – and the active support of – the company's expertise on work environment issues, which would act as a kind of interactive and guiding resource when requested by the production departments.

In short, the strategy applied by the engineering company might be described as continually introducing new participative methods to deal with risks at the workplace. This strategy guarantees that attention is paid to safety, while at the same time workers' involvement forms an intrinsic part of the safety efforts.

Is it possible to have learning with control added?

As just illustrated by the example of the engineering company, approaches that combine learning and control are certainly possible. We suggest that learning as a basic strategy might enable the effective use of in-house occupational services, supported by both expertise and a principle of control. These occupational services would guarantee "safety thinking" among production personnel by ensuring that they recurrently have different kinds of tasks to perform that directly address risk issues (such as following a machine safety checklist, reporting incidents for a week and so on). Such a measure would provide a foundation for integrating safety-related thought and processes into everyday work at all levels within the organization. Since this approach seems possible within an overriding philosophy of control – specifically control with learning added – then relinquishing control as an overriding philosophy and instead introducing *learning with control added* would offer a possible route to attaining safety. In a case such as the special department in the newspaper printing plant discussed earlier, a successful approach along these lines might be to use safety experts who – on the basis of pre-negotiated agreements – fed control-oriented tasks into regular work practices. Such an approach would have the advantage of keeping safety as the responsibility of a particular person, while at the same time seeing to it that safety also became the concern of everyone involved.

In May 2001, twenty-four safety engineers in workplaces[2] from a variety of industrial sectors in Sweden responded in a short questionnaire concerning the conflicting principles we have presented above. The questionnaire was distributed to the engineers while they were attending a specialist training course. The majority (eighteen) of responses indicate that there is coherence between the approaches by which operations for production and safety are run in the various workplaces. But there are also cases (six) of discrepancy between approaches. All but four of the twenty-four engineers provide examples of situations in which safety requires "self-governed thinking, to act, take initiatives, and experiment," and sixteen engineers state that "there is a need for this type of safety approach to increase." Also, thirteen respondents acknowledge that there are conflicts between the principles, that guide risk control on the one hand, and safety learning on the other. Most of these respondents (ten) provided specific examples of these conflicts.

The respondents provided various examples of situations in which production-oriented initiatives and the need for creative improvements to production processes resulted in insufficient attention being devoted to safety. Safety ran the risk of being totally forgotten. The engineers' examples of situations in which safety instructions, for example, were forgotten when working with tasks requiring innovative thinking included maintenance work, developing production through improvisation and the offering of creative solutions to production disturbances.

Conclusion

In this chapter, we have addressed the problem of reconciling two potentially conflicting approaches to risk management – namely risk control and safety learning (Backström 2000) – and the possibilities that these approaches might coexist at the workplace. Focusing on the continually ongoing processes whereby risk and safety are socially constructed, we adopt the stance that risk and safety constitute phenomena which are constructed in everyday work processes. This perspective leads to the insight that through risk management it is possible to consciously guide and govern the processes of constructing safety.

As our point of departure, we took the normative position that managing risks through both control and learning is necessary in order to achieve and maintain a safe work environment. It is also assumed, however, that there are fundamental conflicts – even incompatibilities – between the two managerial approaches. Based on our observations and fieldwork, we concluded that both approaches should be considered independently as well as in relation to each other. In each case, the issue of the role of culture in encouraging and maintaining safety has to be dealt with. We argue that although the two concepts of control and learning imply two opposing and conflicting cultures, it is still necessary to work with them simultaneously and allow their co-existence.

Our empirical example of an engineering company that combined control and learning illustrates that successful accident prevention can be achieved when participation and learning are added to a traditional safety culture that is based on internal control. Whether or not the opposite type of situation – that is, where the controlling aspects of safety work have to function within an approach based on learning – would be successful is, however, still an open question. We have as yet no definitive answer as to whether the principles of control and learning can coexist and thrive together under such circumstances, or whether their basic assumptions are too much in conflict. Based upon the theoretical and empirical ideas presented in this chapter, one possible approach would be to stress the importance of identifying learning and control as different logics. At the workplace level, the challenge would then be to design ongoing means of consciously stimulating their (separate) development, as well as nurturing (necessary) interactions and transactions between them.

Notes

1   This was previously the National Board for Occupational Safety and Health (NBOSH) and the Labor Inspectorate.
2   The size of these workplaces varied from just a few employees to 7,000 (where sixteen workplaces had 100 employees or more and seven workplaces employed 1,400 or more).

References

Alvesson, M. (1990; 3rd edn 1997) *Kommunikation, makt och organisation*, Stockholm: Norstedts Ekonomi.
Backström, T. (1998) *Internkontroll vid DNEX Tryckeriet AB – dess funktion och framväxt*, Arbetslivsrapport No. 1998: 10. Arbetslivsinstitutet.

—— (2000) "Control and learning in safety management," in D. Podgórski and W. Karwowski (eds) *Ergonomics and Safety for Global Business Quality and Productivity*, Warsaw, Poland: Central Institute for Labour Protection, Warsaw.

Backström, T. and Döös, M. (1995) "A Comparative Study of Occupational Accidents in Industries with Advanced Manufacturing Technology," *International Journal of Human Factors in Manufacturing* 5(3): 267–282.

—— (2000) "Management of Safety During Correction Work in Automated Production," paper presented at 26th International Congress on Occupational Health (ICOH), Singapore, August 2–September 1, 2000.

Berger, P. and Luckmann, T. (1966) *The Social Construction of Reality. A Treatise in the Sociology of Knowledge*. Harmondsworth: Penguin Books.

Cooper, D. (1998) *Improving Safety Culture – A Practical Guide*, Chichester: John Wiley and Sons.

Coyle, I.R., Sleeman, S.D. and Adams, N. (1995) "Safety Climate," *Journal of Safety Research* 26(4): 247–254.

DeJoy, D.M., Murphy, L.R. and Gershon, R.M. (1995) "Safety Climate in Health Care Settings," in A.C. Bittner and P.C. Champney (eds) *Papers Presented at the Advances in Industrial Ergonomics and Safety VII*, London: Taylor and Francis.

Dixon, N. (1994) *The Organizational Learning Cycle. How we can Learn Collectively*, London: McGraw-Hill.

Douglas, M. (1992) *Risk and Blame. Essays in Cultural Theory*, London: Routledge.

Döös, M. (1997) *The Qualifying Experience. Learning from Disturbances in Relation to Automated Production*, Dissertation, Department of Education, Stockholm University. (In Swedish with English summary) Arbete och Hälsa 1997: 10, Solna: National Institute for Working Life.

Döös, M. and Ohlsson, J. (1999) "Relating Theory Construction to Practice Development – Some Contextual Didactic Reflections," in J. Ohlsson and M. Döös (eds) *Pedagogic Interventions as Conditions for Learning – The Relation Between Theory and Practice in Some Contextual Didactic Examples*. Stockholm: Department of Education, Stockholm University.

Döös, M., Wilhelmson, L. and Backlund, T. (2001) "Kollektivt lärande på individualistiskt vis – ett lärdilemma för praktik och teori," in T. Backlund, H. Hansson and C. Thunborg (eds) *Lärdilemman i arbetslivet*, Lund: Studentlitteratur.

Dwyer, T. (1991) *Life and Death at Work. Industrial Accidents as a Case of Socially Produced Error*, New York and London: Plenum Press.

Edelman, B. (1997) *Shunters at Work – Creating a World in a Railway Yard*, Dissertation. Stockholm: Stockholm University.

Elgstrand, K. (2001) "Swedish Initiatives in International development in Occupational Health," *International Journal of Occupational and Environmental Health* 7(2): 127–129.

Hart, H. (1999) "Ständiga förbättringar som komponent i en ledningsstrategi för förändring," in T. Nilsson (ed.) *Ständig förbättring – om utveckling av arbete och kvalitet*, Solna: Arbetslivsinstitutet.

Hart, H., Berger, A. and Lindberg, P. (1996) *Ständiga förbättringar – ännu ett verktyg eller en del av arbetet i målstyrda grupper?*, Solna: Arbetslivsinstitutet.

Kjellén, U. (2000) *Prevention of Accidents Through Experience Feedback*, London: Taylor and Francis.

Kolb, D. (1984) *Experiential Learning. Experience as the Source of Learning and Development*, Englewood Cliffs, NJ: Prentice-Hall.

Lindheim, C. and Lien, K.M. (1997) "Operator Support Systems for New Kinds of Process Operation Work," *Computers and Chemical Engineering* 21(Suppl.): 113–118.

Löfberg, A. (1989) "Learning and Educational Intervention from a Constructivist Point of View: The Case of Workplace Learning," in H. Leymann and H. Kornbluh (eds) *Socializaion and Learning at Work. A New Approach to the Learning Process in the Workplace and Society*, Aldershot: Avebury.

—— "Nonformal Education and the Design of Workplaces as Learning Contexts," paper presented at International Conference on Learning and Research in Working Life, Steyr, Austria, 1–4 July 1996.

Mezirow, J. (ed.) (1990) *Fostering Critical Reflection in Adulthood*, San Francisco: Jossey-Bass.

NBOSH (1996) *Internal Control of the Working Environment* Ordinance No. 1996: 6. Available HTTP: <http://www.av.se/english/legislation/afs/eng9606.pdf> (accessed August 2002).

Ohlsson, J. (1996) *Collective Learning. Learning within Workgroups at Daycare and Leisure Centres for Children*, Dissertation (in Swedish with English summary), Department of Education, Stockholm University.

Ortner, S. (1984) "Theory in Anthropology Since the Sixties," *Comparative Studies in Society and History* 26: 126–166.

Piaget, J. (1970) *Genetic Epistemology*, New York: Columbia University Press.

Reason, J. (1997) *Managing the Risks of Organizational Accidents*, Aldershot: Ashgate.

Sitkin, S.B., Sutcliffe, K.M. and Schroeder, R.G. (1994) "Distinguishing Control from Learning in Total Quality Management: A Contingency Perspective," *Academy of Management Review* 19(3): 537–564.

Snell, S.A. and Dean, J.W. (1992) "Integrated Manufacturing and Human Resource Management: A Human Capital Perspective," *Academy of Management Journal* 35: 647–704.

Watson, J.L. (1977) "Introduction: Immigration, Ethnicity, and Class in Britain," in J.L. Watson (ed.) *Between Two Cultures. Migrants and Minorities in Britain*, Oxford: Basil Blackwell.

Wilhelmson, L. (1998) *Learning Dialogue. Discourse Patterns, Perspective Change and Learning in Small Group Conversation*, Dissertation (in Swedish with English summary) Arbete och Hälsa 1998: 16, Solna: National Institute for Working Life.

Williamson, A.M., Feyer, A.-M., Cairns, D. and Biancotti, D. (1997) "The Development of a Measure of Safety Climate: The Role of Safety perceptions and Attitudes," *Safety Science* 25: 1–3, 15–27.

Zohar, D. (1980) "Safety Climate on Industrial Organizations: Theoretical and Applied Implications," *Journal of Applied Psychology* 65(1): 96–102.

Index